Completing your Thesis

A PRACTICAL GUIDE

Nelleke Bak

Van Schaik
PUBLISHERS

Published by Van Schaik Publishers
1064 Arcadia Street, Hatfield, Pretoria
All rights reserved
Copyright © 2004 Van Schaik Publishers

First edition 2004
Second impression 2005

ISBN 0 627 02557 9

Commissioning editor Leanne Martini
Production manager Ernst Schlatter
Editorial coordinator Julia Read
Copy editor Marion Marchand
Proofreader Judith Marsden
Cover design by GL Interactive, Pretoria
Typeset in 9.7 pt on 12.5 pt Utopia Regular by Icon Agency, Clubview
Printed and bound by Paarl Print, Oosterland Street, Dal Josafat, Paarl, South Africa

Every effort has been made to obtain copyright permission for material used in this book.
Please contact the publisher with any queries in this regard.

Please note that reference to one gender includes reference to the other gender.

ACKNOWLEDGEMENTS

The ideas in this book have developed over the last fifteen years during my collaborative involvement with colleagues and Master's and Doctoral students. What started off as a list of "some study suggestions" grew into ever more extensive "handy hints" and later into a series of generic booklets aimed specifically at assisting students to finish their theses. Countless people have added to the ideas in this book in countless ways, so much so that the boundaries of authorship over the years have become blurred. I apologise to those whom I ought to have acknowledged properly, but haven't, and apologise to those whose ideas I've adapted in ways that might have changed their initial thrust. Many thanks to all of you for your contributions.

But there is one significant author whose voice resonates throughout these pages. Wally Morrow's combination of practical advice, intellectual challenge and pedagogical insight produced the initial Thesis Guide for our small group of Master's students. His example is the foundation on which this book is based, and my deep thanks go to him.

Nelleke Bak
Cape Town
October 2003

CONTENTS

Appendices 107

PREFACE

How does one structure a guide on a process that is often not linear, is difficult to define in terms of clear beginnings and end points, and is characterised by discipline-specific conventions? Writing a generic guide that will assist students from a variety of disciplines entails steering a path between being too general (and thus not really helpful) and being too discipline- or institution-specific (and thus not addressing the concerns of others). I hope that this guide succeeds in keeping a productive balance between offering generic guidance and illustrating with subject-specific detail, and that its practical advice will help you to complete your thesis.

However, the process of writing a thesis is seldom a clear, systematic, step-by-step progression towards a finishing point. More often than not, the process doubles back on itself, surges ahead, stalls, goes sideways, but eventually does reach completion. There might be times when you have to revisit your problem question, change your research methods, or adapt your aims in the light of unanticipated developments or insights. Also, reading and writing are reciprocal processes – writing directs your reading, and reading redirects your writing. Despite these dynamics, there are systematic broad steps you can follow in writing your thesis.

The book is organised in such a way that it roughly follows the main stages of thesis production. Part A focuses on how to develop a sound research proposal; Part B deals with how to negotiate and sustain the supervision relationship and the actual thesis writing process; and Part C addresses the requirements for submitting the thesis for examination. Since the process of thesis writing is more spiral than linear, there are many cross-references between the three parts.

Institutions often have different names for aspects of the research undertaking. For example, at Master's level the research component is sometimes called a "minithesis", or a "thesis of limited scope", or a "research project", or a "research essay", or a "full thesis", and at Doctoral level some institutions refer to a "dissertation" in place of a "thesis". Some institutions distinguish between a "supervisor" at Master's level and a "promoter" at Doctoral level, whereas other institutions don't. In order to avoid these differences in terms, I will use the term "thesis" to cover both Master's and Doctoral writings. Where it is necessary to distinguish between these, the terms "Master's minithesis", "Master's full thesis" or "Doctoral thesis" will be used. And for the sake of simplicity, I shall use the term "supervisor" for both Master's and Doctoral theses.

One of the thrusts of the book is to signal to students that doing research

Van Schaik Publishers

doesn't necessarily mean doing empirical research – many students seem to think that they *have* to interview people or administer questionnaires if they are to do research. So, in Part A – when discussing chapter outline – I spend quite a while discussing purely theoretical studies, as well as empirical studies.

There are many different books and sources of information on doing Master's and Doctoral research. But remember, reading about how to do it is no substitute for actually doing it. You will really learn to write academically and learn to do research only by *doing* it.

Learning to write a thesis is like learning to ride a bicycle – you can learn a little about how to do it by listening to others telling you; you can even learn a little by watching others do it; but ultimately you only learn to do it by doing it yourself, by getting on the bicycle, falling off, getting back on, and proceeding with a little more confidence a little further each day. This book, I hope, will serve as a steady hand on the saddle as you set off on your thesis ride.

Nelleke Bak
October 2003

Part A

Developing and submitting
a research proposal

Part A aims to help you develop a sound proposal, one that will enable you to write your thesis in a focused and disciplined way. Research proposals, however, differ from subject area to subject area. There is no single format, but there are some central features that all proposals share. Part A of this book presents a generic research proposal guide that you can consult in the absence of any other specific directives on developing the proposal. The key components of a proposal are these:

- A description of the research problem
- An argument as to why the problem is interesting and important
- The debates within the academic literature relevant to the research problem
- A description of the way in which the problem will be approached
- The proposed research methodology.

See Appendix 1 for an outline of what a thesis is and what it is not.

The research proposal is a fundamental part of the process of thesis production. Without a clear proposal, it is unlikely that you will be able to embark on a systematic investigation and discussion of a problematic issue in your area of research. The initial step of the process starts with an idea of what you would like to investigate, based on academic debates on the subject. This idea is then formulated into a research problem, usually in the form of a question or a hypothesis. The procedure you propose to follow in order to help you answer the problem question, or test your hypothesis, is your research design. All this you write up in your research proposal. So, you see, the proposal already takes you a significant way towards developing your thesis. Once your research proposal has been finalised, you start on your actual thesis. It is worth investing time and effort into developing a sound proposal – the more rigorous the proposal, the easier the writing of the thesis. Of course, a proposal cannot anticipate all the findings or conclusions you may arrive at in your investigation, but a clear map through the terrain of the research area will prevent you from losing your way in the tangled field.

Thinking about the research proposal

1.1 WHAT IS A RESEARCH PROPOSAL?

Research is an activity that is essentially in the public domain. The kinds of issues that are being researched, the way in which they are addressed and investigated, and the dissemination of the findings, are all embedded in shared practices of the discipline in which you will be working. Becoming a researcher is like joining an academic conversation. You listen to what the topic of conversation is, you note who responds to whom, you become aware of certain groupings and dominant trends, and you observe certain similarities and differences that emerge. You are part of a community of "discussants", sharing ideas, building on what others have said, replicating findings, asking questions and contributing to the public debate. It is therefore appropriate that your proposed research is laid out for inspection and comment by others in the academic community.

The research proposal is a planning document that outlines your thinking about a research problem and describes what is to be studied and how. The thesis writing is a demanding project for which you need to plan. Without proper planning, it is likely that your reading will lack direction, your writing will lack focus and your data collection will run into major problems. Of course, having a research proposal is no guarantee that you will avoid these dangers, but having a clear proposal will certainly minimise wasted time.

The main challenges you will be faced with in writing your proposal are:

- To move from a research idea to a research problem
- To gain clarity on the unit of analysis
- To select an appropriate research design
- To conform to the style and format of a proposal.

1.2 WHY DO I NEED A RESEARCH PROPOSAL?

Imagine that you are planning a journey by car from Cape Town to Polokwane. Before you leave, you make certain arrangements: buy a map, plan the route, fill up the car, organise someone to feed the cat while you are away, telephone friends in Bloemfontein to arrange for your overnight stay, draw money from the bank,

© Van Schaik Publishers

pack food for the trip, and so on. Just so with the research journey. The proposal is your map, your itinerary, the preparations you've undertaken and a list of necessary provisions for the journey. The research proposal is a way of thinking clearly about the destination you want to get to, the means of getting there and the arrangements you have to make in order to undertake the research. Without such a proposal, you might easily get lost.

A research proposal is a useful document to you, your supervisor, your funders and the broader research community. For **you, the student**, it is helpful in that it outlines your thinking about what you will be investigating – the focus, the limits, the logical development of your investigation and the methods you will be using. The better your planning, the better your research undertaking will be organised. Through the development of the proposal, you come to a clearer understanding of the debates in the literature, the main considerations of the issue, the potential pitfalls, the perspective from which to approach your research, and the ways in which you will gather information from available sources.

For your **supervisor**, a proposal is an indication of whether you have done adequate thinking about the topic and made sufficient preparation for the study. It also gives your supervisor an indication of your ability to put your ideas into clear and logical writing. A proposal forms the basis of a kind of contract between you and your supervisor. It gives you and your supervisor a plan of action to follow in order to reach completion of the thesis (your ultimate aim!).

For **funders** the proposal gives a clear indication of whether it is something feasible and worth supporting. And for the rest of the **academic community**, it is an indication of the focus of your investigation and how it links to the ongoing debates in the literature.

> The research proposal is your opportunity to persuade the academic community that you know what you are talking about. That you have thought through the issues involved and are going to deliver. That it is worthwhile to take the risk and give you licence to get on with it.
>
> (Robson 1994: Appendix A)

Remember that you are the one who will invest most in this project. You are the one who has most to lose if you get into a situation where the research is a failure or you find that you have wasted enormous amounts of time going in the wrong direction. You want the best insurance policy you can get. So it is worth investing time and effort in developing a good research proposal.

A good research proposal will help you to

- define and formulate your research question
- narrow down the study to a manageable form within the prescribed time limits
- structure your writing and the development of the overall argument
- avoid wasting time in the literature search and data collection stages of the project.

1.3 ON WHAT CRITERIA ARE RESEARCH PROPOSALS JUDGED?

Bear in mind that different supervisors and examiners look for different things. Discuss with your supervisor the main points that you should develop clearly in your proposal. However, in general the following questions are usually considered in the examination of proposals:

- Does your proposal have focus?
- Is it a topic worthy of academic study?
- Do you have a clear idea of what you plan to research? Do you demonstrate an adequate understanding of the debates in the literature on this topic?
- Is the project feasible? Do you have a realistic idea of how you are going to tackle the investigation?
- Is it manageable within the time constraints?
- Do the bibliography and referencing conform to accepted conventions? Are they technically faultless?

See Appendix 2 for a checklist for evaluating research proposals.

1.4 WHEN AND WHERE ARE RESEARCH PROPOSALS SUBMITTED?

Different universities and faculties have different requirements, so check with your supervisor or the chairperson of the faculty committee responsible for postgraduate studies.

If you are doing a Master's by coursework and minithesis, you will normally be required to submit your proposal some time during the first year, after you have completed the coursework. If you are applying for a Master's or Doctorate by full thesis, contact the specific department. Your interest area will be discussed and, in most cases, a potential supervisor will be identified. Working with the supervisor, you will have to prepare a research proposal. This will be tabled at the relevant faculty committee. The committee has the following options:

- It accepts the proposal as adequate.
- It refers the proposal back for reworking and resubmission.

Your thesis title will be registered for the permitted number of years (usually two years for Master's and four years for Doctorate). Most universities allow students to apply for special permission to extend their studies beyond the set number of allotted years.

1.5 WHAT IS THE FORMAT AND LENGTH OF A RESEARCH PROPOSAL?

Again, requirements differ from university to university. Check with the relevant person (your supervisor or the chairperson of the faculty's committee responsible for postgraduate studies). Here is a comprehensive list, in the required sequence, of the various section headings:

- Cover page
- Abstract (on separate page)
- Title
- Ten keywords (or composite words)
- Aims of the research
- Rationale
- Framework of the research/literature review
- Research problem/hypothesis
- Delimitation of study area/assumptions on which the research rests
- Interpretation of key terms
- Research design
- Research methods
- Ethics statement
- Chapter outline
- Time line
- Budget
- Dissemination of research
- Preliminary bibliography

Sometimes sections are combined. For example, research design and research methods are often combined when the proposed research is a theoretical study. Sometimes sections are left out. For example, if the research will not involve empirical investigation of people or animals, an ethics statement is usually not necessary. Chapter 2 of this book will deal with each specific section of the proposal.

See Appendix 3 for examples of well-written proposals.

The length requirement of the proposal also differs from institution to institution. *Master's thesis proposals* can vary from 5–15 pages (excluding cover page), one and a half-spaced typing. *Doctoral thesis proposals* can vary from 10–40 pages, one and a half-spaced.

· ·

1.6 HOW DO I DECIDE ON A TOPIC?

The formulation of an appropriate and interesting research topic is perhaps the most demanding and difficult part of your proposal development. So don't despair if you don't have a clear topic from the start. Getting focus and clarity is part of the proposal writing process. Moreover, the process of developing your proposal is often not a linear one, moving logically from phase to phase – it is not a mechanical process with clear steps to be followed rigidly. More often than not, it is a process that turns back on itself, surges ahead in leaps, collapses after some more reading, and finally reaches greater focus. The process is more like a spiral than a straight line. However, that doesn't mean that you should avoid planning the process systematically. In this section the process of developing a research proposal is explained on the basis of a number of logical phases. In general, the typical phases include the following (Conradie 2000: 1):

- Choosing a particular research topic
- Identifying and formulating the specific research problem
- Conceptualising the process of investigating this problem
- Collecting, analysing and interpreting the relevant data to investigate the problem
- Writing the final proposal.

Often students do not know what they want to write on. If you don't have a clear idea, asking you to "choose" any topic that interests you is to leave you feeling quite lost and under pressure to find something, anything! Such looseness can result in you floundering for a long time (not altogether a bad thing in itself perhaps!), and often leads to ambitious, vague, uncritical approaches to a topic. In cases where you clearly do not know what to write on, you can ask your supervisor to be quite interventionist with suggestions of structured topics – this can be a strategy for *strengthening*, not silencing, struggling students.

See the process of developing your research proposal as one of increasing refinement and focus. The first step, then, is to identify an appropriate and interesting theme, a broad area of study. Ask yourself why this area interests you, why it is a relevant one to investigate, and how this theme has been addressed in the literature. Once you have identified a likely theme, focus on a particular topic in that theme. Ask yourself what specific issue within that broad theme you want to investigate. Why do you want to investigate it? Then, once you have identified the topic, focus on a *problem* within that topic.

1.6.1 How do I start to find a topic?

Think of your thesis as trying to address a problem, a gap in the literature, a puzzle, a muddle, an ambiguity or a tension. Writing about something that is straightforward and unproblematic doesn't constitute an *investigation*. Mere description is not research. But not all problems are viable topics for academic research. Ensure that the problem you choose to investigate is part of an ongoing academic conversation, one that has been addressed in some way or other in the literature. (This is especially important in the case of Master's theses, where you are not required to make an original contribution to the field of knowledge. You are required mainly to demonstrate your familiarity with the academic debates on a particular issue.)

The following may be fruitful places to start looking for an appropriate theme:

The coursework

(This is applicable in the case of a Master's degree by coursework and minithesis.) The coursework part addresses issues in your field of study that are relevant, interesting and usually contested (that is why they are chosen for inclusion in the coursework). Think of some of the class debates and assignment topics you were engaged in. Are there any of these that you would like to pursue? You might place them in a different context, and look at possible implications thereof. The advantage of choosing a theme from the coursework is that you will have a set of readings ready to start off with, will have written something on the theme (maybe an assignment that could be adapted for inclusion in the minithesis), and will have engaged in discussion that will have helped sharpen your thinking about it. If you are just starting with the coursework and will be tackling the thesis part only later on, begin thinking about a possible topic from the start of your studies. Get a notebook and jot down any interesting topics that surface during class discussion and ask your lecturer how such a topic might be developed for research in a thesis. By the end of your coursework, you will have a rich resource of possible topics to draw on.

The literature

Think of an academic book or journal that you have read. What was the title? What issue did the book or article address? Can you place the issue in a different context? For example, could you place the issue in a different time period, or a different place (such as in post-1994 South Africa), or look at the issue as it pertains to a specific group (such as women or children), or address the issue from a different theoretical perspective (such as focusing on power relations within that issue)? Go to the library and browse through the contents pages of journals in your subject area. Look at the kinds of issues the articles are addressing. Is there a theme or issue that you want to pursue? Also, ask your supervisor about successful theses that previous students have written. What issues did these theses address? Could you pursue one of them? The advantage of choosing a theme from the literature (including previous theses) is that you have a reading list at hand – the references listed in the bibliography – on which to build your investigation. Don't think because someone has already written about the topic that you can't do so – philosophers have been writing on "free will" for centuries and, I suspect, will continue to do so for a long time yet. There is seldom a definitive answer to a problematic issue. Your tackling the issue might build upon what others have previously written or refine some of their ideas (and so you contribute to the stock of knowledge).

8

An existing research project

Many supervisors are engaged in existing research projects that have already been registered. Find out who in the department is involved in what kinds of research projects. Joining such a project will give you a solid base and framework (and possible funding) for your research. The advantage of this is that you feel part of a research group, sharing resources, readings and findings, speaking the same academic discourse, as well as receiving support and guidance from group members. Funded research proposals have a pool of money that could assist you with your research. If the supervisor has a National Research Foundation grant, there may be some available student bursaries attached to it. And – an important point – registered research projects have deadlines that serve to motivate you to produce on time.

Find out what other students are working on

Being part of a group of students who are all writing on the same topic can be very stimulating. It can serve to highlight the excitement of flexible and creative academic thinking, tackling the same broad topic from different perspectives. It also creates the sense of a research "group", wherein members support one another and share readings. A group serves to create a sense of ongoing energy and direction. This is extremely helpful for those inevitable times when your motivation flags and focus blurs. A "research group" of postgraduate students is more prevalent in the natural sciences, but there is no reason why it should not be encouraged in the social sciences and arts as well. So, if there are enough students who are interested in writing on the same topic (with different emphases and different foci), you might look to establish such a group with your fellow students and with the assistance of your lecturers.

Ask academics in the discipline

Academics in the discipline have a good idea of what interesting topics there are in the field. Maybe some of their colleagues elsewhere are working on exciting projects; maybe they have read a book or article and thought the topic stimulating; maybe they have just returned from a conference and have some good ideas for you to pursue. Academic work is collegial, so ask your lecturers for suggestions. Attend conferences yourself if you can, and interact with as many attendees as possible. Network for suggestions and advice.

Your personal experience

This is a tricky source from which to choose a topic. It is often a source of great passion and interest, which might help to motivate you. But ensure that the issue you choose from this source is a viable, academically researchable one. The disadvantage of choosing a topic from personal and workplace experience is that the problem is frequently a pragmatic rather than an academic one. Moreover, suitable literature is often difficult to find.

A PhD programme should really be a marriage of your and your supervisor's interests. However, it is important not to confuse your supervisor's enthusiasm for a research topic with your own academic interests.

1.6.2 How do I identify a problem?

Remember, the identified topic within the broader theme must focus on a problem. Your thesis will try to address that problem through a systematic, disciplined discussion, informed by the literature. Not all "problems" are researchable. For example, the government may have the "problem" of insufficient funds to implement a policy of low-cost housing. The solution to this "problem" would simply be: more money! But there may be all sorts of other researchable problems underlying this issue: Should the government cut back on health provision in order to provide housing? What are the tensions around budget constraints? Should housing be the government's priority? Could the provision of housing be privatised? How can low-cost housing promote economic justice? The answers to such questions are not obvious and are often highly contested. So these kinds of question address problematic issues and are therefore suitable for academic investigation.

One of the most significant ways of finding a researchable problem is to *read, read, read*. There is no short cut. Remember, you are entering an ongoing academic conversation and you need to demonstrate that you are familiar with the academic debates pertaining to the particular issue you have chosen to address. These debates comprise the literature in the field. Furthermore, it is through engaging with these debates that you identify what the substantive problems are, particularly the problems that are, as yet, unresolved. Reading about the theme and topic sharpens your thinking and refines your research problem. It is worth spending time on formulating a clear, focused research problem – it will save you lots of time and frustration later on.

There are different kinds of problem that may become the focus of your thesis:

- **Conceptual:** your thesis may address and evaluate different interpretations of key concepts. Or it may analyse the meaning of a concept, such as "human rights", and investigate its relationship to other concepts such as "civic responsibility" and "freedom". Or it may investigate the implications of a concept, such as "democracy", in a particular context, such as education. Or it may compare and contrast different models developed in the literature, such as different models of management for small businesses.

- **Epistemological or logical**: your thesis may address some problem in thinking. It may analyse the validity of arguments that support a particular position. Or it may investigate some contradictions or paradoxes in thinking, for example: Is self-deception possible? Can one know and not know something at the same time?

- **Exegetical**: your thesis may address a semantic issue. It may involve translations, semantic studies and literary analysis. For example, it might compare different translations of a religious text or of a novel.

- **Social, political or economic**: your thesis may address a problem such as community involvement in decision-making; the role of unions in a free market economy; the implications of HIV/Aids for family structures. Remember, not all problems you read about in the newspaper or encounter in the workplace or your personal life are researchable problems. Look for the underlying interesting questions to ask.

- **Ethical**: your thesis may address what would be the most honourable or appropriate course of action in a particular situation. Or it may investigate what some of our moral stances are, or our views on human well-being that drive certain policies, for example your thesis could ask why we should teach our children to care for the environment. Or it may investigate the moral dilemma of health caregivers, for example whether to treat information on their patients' health status as confidential while they might pose a threat to the well-being of others. A word of caution, though, a thesis is a piece of research, not a sermon or ideological tract!

- **Legal, policy**: your thesis may give a clear exposition of what are often difficult legal tracts, and analyse some of the underlying assumptions. Or it may look at some of the problems of policy implementation, such as the enforcement of secondary rights. Or it may examine the justification for particular policies, for example, the justification for the policy of intellectual property; or some of the implications or consequences (intended and unintended) of a policy, such as the economic consequences of the privatisation of a particular sector.

- **Theoretical**: your thesis may compare and contrast different views expressed in the academic literature on a particular issue, say the interpretation and implementation of affirmative action employment policies. Or it may evaluate the arguments that support or reject a particular position, such as the link between HIV and Aids. Or it may examine how a theory needs to be modified if placed in a different context.

- **Historical**: your thesis may examine gaps in historical narratives, or trace the development of a particular issue over time, such as the changing role and nature of labour unions. Or it may trace the different interpretations of a concept such as "citizenship" in political studies or "child" in law over a historical period.

- **Empirical**: your thesis may try to address the problem of lack of information on a particular issue through doing fieldwork or experiments, in addition to engaging with the relevant literature on the issue. Your thesis may want to replicate an empirical study or experiment done elsewhere, for instance to determine the rate of illiteracy levels among taxi drivers in Bloemfontein, based on a similar study done in Cairo, Egypt. Or you may set up an experiment to test a specific hypothesis. Or you may want to look at the actual implications of a particular policy, perhaps looking at investment strategies used by black women who run small businesses.

Your thesis may address a combination of these problems. However, remember that a *focused* problem will give your thesis coherence. Too many different kinds of problems can lead to you losing the thread of your research.

1.6.3 I have identified a topic and a problem within it. What now?

The proposal writing process, as noted earlier, often does not follow a systematic linear development. Again, *read, read, read*. You might find that the more you read, the more lost and unsure you become. Don't despair. It is part of the clarification process. Through reading, you start to gain focus not only of what to *include*, but also of what to *exclude*. So much has been written. You cannot possibly include it all, so you must make informed decisions about what you are *not* going to investigate: you need to decide on the limits of your study. Debates are usually not neatly separated into clearly defined fields. You might find that in order to address a particular issue, you need to address something prior to this. But in order to do *that*, you need to tackle another problem that has surfaced, and so on and so on. Where does it stop? It doesn't. At least, not in a neat end point. *You* have to decide where your starting and end points are within this ongoing academic conversation of which you are part. Through reading, you modify your topic, sharpen it more, and give it greater focus. No one expects you to tackle the whole field, but what you do need to do is to state clearly what your starting assumptions are, as well as the boundaries of your investigation.

See section 2.8 on the research problem/ research hypothesis.

Once you have clearly articulated your problem question (which can also be expressed as a hypothesis), you need to think about *how* you are going to research it. Read about different ways in which a problem can be investigated (or a hypothesis tested). You might find that your problem is "not feasible" – it just cannot be researched, either because of a logical difficulty or because of practical reasons. In this case, you'll have to refine and rework your problem question. For example, impact studies or examining the "effect of x on y" are notoriously difficult to research because of the complexity of causality, which is hardly ever linear or simple. Or you might find that you cannot get access to particular historical documents central to your thesis, or the people whom you were planning to interview are just too difficult to access. Again, rethink your thesis problem.

Choosing a topic, identifying a problem and conceptualising the process of investigation are phases that repeat themselves within the spiral development of your proposal. A suggestion: keep a notebook in which you jot down all your ideas and the development of your thinking. It will trace the process that at times might seem to you to be going nowhere, but if you look at how your thinking has progressed in the notebook, you might take heart and be reminded of how far down the clarification road you have already come.

1.6.4 How do I refine the topic and get a clearer focus?

The vast majority of proposals start off by being too ambitious, and therefore not manageable. Many students fear that they may not have enough to say if their topic is too narrow. This fear is usually based on not really understanding the complexity of the issue at hand. Rather investigate a small topic in depth, than a large topic superficially. So general advice is to focus the topic on a particular problematic issue.

Think of your thesis as developing a picture of a particular issue, but instead of using paints and brushes, or film and camera, you use words, numbers, tables and graphs. As with most photographs, some things in your thesis are in the fuzzy background, whereas others are sharply in focus in the foreground. Decide which aspects of your topic belong in the "fuzzy background" and which aspects in the sharply focused area. The "fuzzy background" will typically sketch the context in which your topic is situated, some general considerations about it, and the trends in the "academic conversation" thus far. This part of your thesis you may treat with broad brushstrokes, in other words you need not go into fine detail here – you may state your starting assumptions, and you may cite literature that develops the arguments for certain positions without you having to reproduce the argument. This broad discussion is a "setting" for the particular aspect you wish to research. In the focus part, you must justify your argument, give critical summaries of the main texts you are drawing on, furnish examples, trace the implications of your position, tease out the details, and so on. Here you are expected to go into detail and to work systematically. That is what gives your thesis focus.

See Chapter 4 and Chapter 6 for more on this.

In summary:
- Get a notebook to jot down ideas for your topic.
- Consider drawing a possible topic from the assignments you've written, class discussions you've had, titles of journal articles or books you've read, current research projects or student research groups in the department, past and current thesis topics, or from personal experience.
- Ask your lecturers to give you suggestions and to direct you to some further readings on topics of possible interest.
- Read for greater focus. Rather investigate a limited issue with rigour, than some broad topic superficially. (Do more investigation of less, rather than less of more.)
- Identify what kind of problem your thesis will address.
- Decide what goes into the "fuzzy background" and what into the sharply focused area.
- Encourage the establishment of a "research forum" of students working on similar issues.

See Appendix 4 for a worksheet on how to choose and refine your topic.

© Van Schaik Publishers

Putting the research proposal together

There are various ways in which proposals can be structured. Often there are differences: between scientific disciplines (Maths and Biology), between scientific cultures (natural and social sciences), and between types of research (empirical and theoretical studies), which are reflected in the structure of the proposal. Check with your supervisor whether there is a specific structure you need to follow. However, most proposals require the following parts in the order given:

2.1 THE COVER PAGE

RESEARCH PROPOSAL

University of _____

Name of candidate: _____

Student number: _____

Proposed degree: _____

Programme/Department: _____

Title of thesis: _____

Supervisor: _____

Co-supervisor (if any): _____

Date: _____

2.2 THE ABSTRACT

Not all universities require an abstract in the research proposal. However, it is a handy item to include, since this short abstract will be taken up into databases listing current research, will enable you to network with other researchers working in the same field, and will give your reader a clear idea of the main thrust of your intended research. In usually no more than 200 words, say what your central problem question (or hypothesis) is, why it is a problem worthy of study, how you will go about studying it, and what conclusions you anticipate. Your abstract should be succinct and informative, giving a reasoned indication of what is intended and why. Although your abstract comes at the beginning of your submitted proposal, it is often the last thing you construct. You can only really write the abstract once you have a clear idea about the topic, problem, justification, research design and main claims. So write these parts of the proposal first, before attempting to put together your abstract.

2.3 THE TITLE

The title should convey clearly and succinctly the topic being researched. It should be brief. Avoid obscure and unnecessarily lengthy titles. Some universities recommend that titles do not exceed 15 words. Start off with a working title, and revisit and reformulate as you read for greater focus. Like a hearty, rich, concentrated stew that has simmered for a long time and from which all excess moisture has evaporated, your title should consist of the essence only.

2.4 KEYWORDS

Provide ten keywords or composite words, which convey what the thesis is about. The keywords should be in a logical sequence: from broad to more specific, or from the central concept to related ones. Some programmes insist on individual *keywords*, others allow *composite words*. Check with your supervisor.

Take care when putting your keywords together – these will be entered into a library catalogue and public database made available to a worldwide research community. Your keywords should reflect, in a kind of telegram style, the main areas or concepts of your thesis, so that someone who does an Internet search will find your title, and someone who reads the keywords can get a fairly clear idea of the focus of your thesis and development of your discussion.

2.5 THE AIMS OF THE RESEARCH

15

Like a journey you wish to undertake in order to get to a specific address, your thesis journey must also have a specific "destination" that you are aiming at, otherwise you can find yourself "driving" around aimlessly for ages and never finishing your thesis. Some academic conventions distinguish between "aims", "objectives" and "outcomes", which generally imply an articulation of what you are planning to *do* (the aims, usually expressed in verbs) and what you hope to *achieve* (the outcomes, usually expressed in nouns). More often than not, for the purposes of a proposal, the aims and outcomes are linked so closely that they all can be in one section.

You must draw a clear distinction between the following two kinds of aims:

- An **academic** aim, which is the issue/problem your thesis hopes to address, based on debates in the academic literature and aimed at an academic audience. Your thesis *must* have an academic aim as its central aim. In other words, it must be an issue that is worthy of *academic* investigation. Formulate your academic aims so that they capture an academic activity. Consider starting your aim/s with words such as: *explore, investigate, analyse, determine, discuss, interpret, understand, demarcate, critique, ascertain, compare, contrast, evaluate, assess.*

- A **strategic** aim *might* follow from your thesis. This is aimed at a non-academic audience, such as policy makers on land distribution issues, health workers in community development projects, managers of businesses. (It is inappropriate for your thesis to have a strategic aim as its central aim.) The strategic aim looks at the possible practical relevance your investigation might have. However, this is always *secondary* to the main academic aim. (Lots of excellent research might not necessarily have immediate and clear pragmatic relevance, for example the study of Homeric metre in ancient Greek poetry, but this doesn't mean that the study is not worthy of investigation.) The strategic aim might start with something like: *improve practice in ..., inform policy in ....*

Avoid having too many aims, which indicates that you don't have a clear idea of your destination. Usually three to four academic aims will suffice.

In summary:
- List your academic aims first, using verbs from the list of examples provided.
- If appropriate, list the strategic aims of your research.
- Ensure that there is tight coherence between your **title**, **keywords** and **aims**.

• •

2.6 RATIONALE/BACKGROUND

You (and your supervisor) will be investing time, energy and resources into this project, so you need to articulate why the study is worth undertaking and why you are interested in pursuing it. In this section you need to explain:

- the **context** that gives rise to your research project. What conditions have led you to propose your research project and to define your aim/s in the way that you have done? (You may hold that certain events, situations, processes and debates require systematic and focused research. You may be of the view that our current knowledge of certain issues is inadequate or that certain issues have been poorly researched. You may be in disagreement with the interpretation advanced by a certain scholar and/or the methodology he used, etc.).

- your **justification** for the research project. What is your interest in the research project? What motivates you to do the project? Why is it worthy of academic investigation? What do you consider to be the significance of the research project? What contribution will the research project make in terms of current knowledge around the issue or problem that is being researched?

The function of this section is to indicate the general importance of the issue you plan to investigate.

• •

2.7 THE LITERATURE REVIEW/THEORETICAL FRAMEWORK OF THE THESIS

Lots of students struggle with this because there are a variety of approaches one can adopt. First of all, bear in mind that writing a thesis is like joining an ongoing academic conversation: academic writers and researchers have probably been "talking" about your issue for a while in journals, books and at conferences. Even if you are addressing a "new" issue, there have been debates that have touched on it in a number of ways over a period. And just as when you join a group of conversationalists, you don't just butt in and interrupt with your own story – you listen first, get an idea of where the conversation is going, who is responding to whom in what way. Once you have a clear idea about this, only then can you speak up and say, "What you said earlier on links up with what X claims", or "The consequences of your conclusion might be the following ...", or "If we were to take your ideas and apply them in a different context, it might have the following implications ...", and so on. In other words, you can only contribute fruitfully to the conversation *once* you have understood what it is about. The academic conversation is a structured one, so it is more apt to call it a discussion. In this section of your proposal, therefore you need to demonstrate to your reader that you have some familiarity with the published academic discussions on your topic.

The framework is vital for guiding the research, for ensuring coherence and for establishing the boundaries of the project. In this section you need to make explicit the interpretation and application of the central issues that will structure (shape and organise) the research. You do this by engaging with the relevant literature. Merely appealing to your own experiences or general knowledge is not enough. For your research proposal, the literature engagement should draw on a limited number of key sources. The thesis itself will expand on the literature.

In this section you will therefore need to indicate:

- What does the **literature in general** reflect about the development of the issue? In what context (historical, geographical, social, intellectual) is most of the literature located? What is the history of your area of study? What are the most recent findings in your area of study? What gaps and contradictions exist among these findings? What new research questions do these findings suggest? Consult a few introductory texts, some standard articles, and chapters in standard works or in topical encyclopaedias in order to sketch an orientation of the kinds of academic debate in the field. Consult some of the major texts and some recent articles to demonstrate that you have a clear sense of the major positions and trends in the field of study.

- What are some of the **specific interpretations** of concepts or issues in the literature you will be drawing on? That is to say, how will you be interpreting some of the key concepts? Also, given the limited scope of a thesis, especially a minithesis, you need to identify some starting points. So what is your point of departure in relation to the *literature* in the field? What are some of the borrowed assumptions you are going to start from? In terms of which interpretations or approaches are you going to analyse your findings? In relation to current knowledge (as reflected in the literature), what do you intend to do? What theoretical model relates to your research topic?

- What **methodological** approaches have been used by researchers on your topic? What results have previous researchers in your field produced? What are the key methodological issues that have been addressed? What are the accepted conventions of research as practised by people within the specific discipline?

Note that your aim/s of the research state the destination of your research activities, whereas the framework is the kind of vehicle in which you will be driving towards your destination. The purpose of this literature engagement section is to establish the theoretical framework for the study, to indicate where the study fits into the broader debates, and thus to justify the significance of your research project against the backdrop of previous research.

2.7.1 Why is it so difficult to construct this part?

Now, let's get back to why so many students struggle with this section. First, it is hard work to engage with the literature (it's much easier telling one's own passionate story, but that is not appropriate for this section). You need to demonstrate your understanding of the main debates in the literature as a basis for developing your own insights later on. Remember, you cannot criticise or adapt something if you don't know, in the first place, what it is you're criticising or adapting!

Second, there's so *much* that has been written. Where does one start and where, if ever, does one end? As noted earlier, it's much more difficult to decide what to exclude, rather than what to include in the thesis. Moreover, everything seems to be connected, so neat divisions are not easily drawn. This is where a focused topic and clear aims are going to help you – skim literature that will be relevant to the fuzzy background, and carefully work through literature that pertains to the focus of your thesis. The more focused your problem, the more directed your reading will be. Here is a hint: go back to your list of ten keywords (which, if appropriately formulated, will capture the main aspects of your research topic). Write them down on a sheet and then, next to each one, jot down two or three authors who address and work with this concept. Then, in your literature engagement, give a brief exposition of how each of those authors has interpreted or used the concept, or what the author's main research findings were on this. Obviously, keywords pertaining to the focused area of the research topic will need more authors or more detailed exposition than the others that belong more in the fuzzy background part.

Third, there are different ways in which one can approach an investigation. Broadly speaking, one can approach an issue from the "inside out" or from the "outside in". (Of course, there are other ways, but let's concentrate on these two prevalent ones.) From the "outside in" generally means that you have identified some large theoretical framework, usually one of the big "isms", and will apply this framework and its attendant tools and interpretations of key concepts to the problem you're planning to investigate. Working with large, scholastic, theoretical frameworks is especially appropriate if your research is on a meta-issue, or if your discipline has well-established models or sets of meanings in terms of which data is collected and interpreted. This approach is graphically illustrated in Figure 2.1.

If you are going to apply some theoretical model, see Appendix 5 for a list of different theoretical positions.

Figure 2.1 Working from the "outside in"

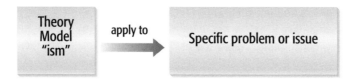

My own bias favours instead an "inside out" approach. My preference for this approach (in the social sciences and arts at least) has its roots in a complicated argument which holds that there are no neat conceptual parcels that correspond accurately with the real, messy, complex world. But this is not the place to embark on a long philosophical debate. Suffice it to say that I think it much more fruitful and accurate to immerse oneself in the practices of the discipline, without worrying too much about how these practices match up with the big theoretical "isms", or which labels one can stick on them. Working from the "inside out" means that, for your theoretical foundations and substantiations, you draw on relevant literature from different authors, maybe even from different paradigms, as the need arises. Of course, keeping coherence within the development of your own argument is important. But with the "inside out" approach, you don't get hung up on labels. And through this, perhaps eclectic, theoretical engagement, a richness of understanding can develop. This approach is graphically illustrated in Figure 2.2.

Figure 2.2 Working from the "inside out"

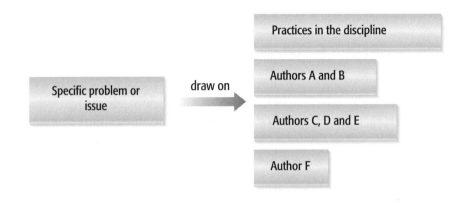

See
Appendix 6
for examples
of a well-
constructed
literature
review and
of a less
satisfactory
one.

In summary:
- Be clear about your topic, key concepts and aims – these will help direct your reading.
- Identify some central texts and discuss how they capture the broad trends in the debates. Note the main arguments, discussions and findings in these texts.
- Identify some important texts that focus on some of your key concepts. Compare and contrast these texts (who agrees with whom, who differs from whom on what points and in which way).
- Discuss what have been the main approaches in the field, i.e. note what research practices characterise the field of study.

2.8 THE RESEARCH PROBLEM/RESEARCH HYPOTHESIS

Arising from your discussion of the literature, you now need to articulate the specific problem that you're planning to investigate (so the research problem comes after your literature review section). This part captures the essential focus of your thesis, and it is therefore important that you spend time on formulating a clear, focused and interesting academic problem that is researchable. This problem is the engine of your thesis – it drives the various sections, directs your discussion towards the destination you want to get to, and informs the way in which you will develop your thesis.

Your central research problem must be:
- **One main** problem, stated in a single sentence, question or hypothesis. If you can't state it this way, it is an indication that you don't as yet have enough focus. You need to demarcate your problem more. I think that formulating

your problem in the form of a **problem question** is fruitful for giving your writing and reading direction – it impels you to *answer* the question. It also assists the coherence of the thesis in that each chapter contributes to the overall answer to the question. Or, in the case of the statement of a hypothesis, this clearly delineates the scope and direction of your study.

- The question must indeed be a **problem**. It should focus on, for example, a gap in the debates, a puzzle, a muddle, an ambiguity, a tension, or a paradox. It can also be a replication of an experiment where further empirical evidence is needed to strengthen (or undermine) the initial findings of the study. You should be able to explain why it is a problem and why it is worthy of study. If the answer to the question is already known, or straightforward, or one on which there is scholarly consensus, then the question is not worth pursuing.

See section 1.6.2.

Your research question must be tightly related to your **research aim/s** and must also emerge from your **research framework**. In other words, the research question or hypothesis must draw on the debates in the literature, and be a contribution to the ongoing discussion. Remember, you are a co-discussant, you're not having a conversation all on your own with yourself!

Pay attention to the **kind** of problem question you pose. Some questions may be of an *empirical* kind in that they seek to obtain information and data that are *descriptive* in nature. Other questions may be of an *analytical* kind, that is, they steer you towards *explaining* a phenomenon. (Questions can start with "Can?", "Should?", "Is?", "How?", "What?", "Why?" etc. Each of these will have a different focus, so make sure you formulate a question that coheres with the aims and title of your research project.)

Here is an example taken from the National Research Foundation (2000: 11–12, reprinted by kind permission of the publishers):

> We start with the idea for a research question ...
>
> *"How have South African mining regulations changed, and what has been the impact of these changes?"*
>
> But this is a huge area! We need to determine some appropriate boundaries in order to make the project manageable.
>
> The broad question needs to be restated more precisely in order to indicate both the purpose of the project and the direction that the research will take. Here it is helpful to define the key terms and concepts that must be investigated, and possibly determine a specific period that the study will cover.
>
> We might do the following:
> - Define safety regulations for purposes of this study as "any Act of the South African parliament which specifically regulates safety on mines within South Africa", so that the project can exclude any regulations imposed by mining companies, mining boards, or particular mine managements.

 Van Schaik Publishers

- Restrict the study to *gold* mining, and to the effect of mining *safety* regulations. After all, there is such a range of different types of mining legislation that the study can't cover them all.
- Restrict the period of time covered by the study. A good starting point would be the Mines and Works Act No. 12 of 1911. Why? Because it was the first piece of legislation passed after the creation of the Union of South Africa in 1910 that controlled general conditions of work on the mines. There could be a number of end points to the study. One could be the Mine Health and Safety Act No. 29 of 1996, which was passed following a Commission of Enquiry into a number of serious accidents in gold and coal mines.

Now we know what type of regulations we are investigating, within which region and sector, and over what time period. Changes in these regulations may have affected many things, however, and we still need to narrow down the areas of impact which the study will consider. Consequences for mining safety? Consequences for staff morale? Consequences for the price of gold? For purposes of this study we might decide the following:

- To look at the consequences for the *cost of gold production*.

In other words, the study will not be concerned with regulations in existence prior to 1911, or to the coal, diamond, or other mining industries, or to changes in management structure, worker morale, or any other aspects in the mining industry beyond the consequences of the legislation for the cost of production.

The research question can now be reformulated as follows:

"What consequences did Acts of Parliament regulating mine safety between 1911 and 1996 have for the cost of gold production in South Africa?"

Note: the initial formulation that was going to look at "impact" has been reformulated to investigating "consequences". This is a much more careful and manageable way of putting it. Impact studies should, in general, be avoided unless you have lots of resources and a clear, limited and easily controlled situation. Much of social life doesn't fit the bill! Impact studies assume a direct, linear causal connection, whereas studies looking at "consequences" or "links" merely hold that there is a connection, but leave space for a much more complex relationship.

2.8.1 Research hypotheses

Research hypotheses are statements of some of the different and possible solutions or responses to the problem question. Some research proposals, especially those planning to undertake experimental work, or planning to do survey studies and empirical studies, need a research hypothesis. You will need to articulate *in proposition (statement) form what you expect to find*. So, spell out the operations and predictions indicated by the hypothesis. Your expectation needs to be based on existing theories, borrowed from other empirical studies or based on logical deduction. Your thesis will aim to test (eventually to verify or falsify)

the hypothesis within the theoretical framework of the research project. Make sure that your hypothesis does not try to investigate too many questions in one research project.

A good hypothesis has several basic characteristics:

- It should be reasonable.
- It should be consistent with available facts and theory, that is, consistent with results established from previous research.
- It should be related to available and accepted techniques.
- It should be testable. You should be able to prove that it is correct or incorrect.
- It should be stated as simply as possible.

> **In summary:**
> - Formulate one main problem. Express this either in question form or as a statement of hypothesis.
> - In trying to find a final formulation, play around with the structure of the problem. For example, see how starting the question with different words, such as "Who", "What", "How", "Why", "Should", "Can", etc., shifts the emphasis of the research focus. Be clear about what focus you want to pursue.
> - Also be clear about what kind of problem your question or hypothesis addresses: conceptual, economic, legal, historical, empirical, etc.
> - Ensure that there is coherence among your problem, title, aims and theoretical framework.

· ·

2.9 DELIMITATION OF STUDY AREA/ASSUMPTIONS

It is important that you state clearly the research assumptions that you will base your research problem/hypothesis on. This assists immensely in avoiding the vague generalisations that so often accompany the lack of proper research planning. A minithesis, in particular, must have carefully specified limits. You cannot tackle everything. Rather than trying to do too much in a superficial way, focus on a small area that you aim to investigate in depth. All research projects must have a starting point and an end point, which necessarily means that you have to set distinct boundaries. This will serve to orientate your reader and to make your study a manageable one.

To delimit your study area requires careful analytical thinking. You are going to highlight certain aspects that are embedded in a seamless web of issues. To do this, you need to break the whole into its constituent parts, and to examine the various elements that make up the whole. Analytical thinking is a sophisticated academic skill that takes practice and systematic untangling. Analysis looks at the various parts, without losing sight of the links between the parts. Your proposal needs to demonstrate that you have been able to demarcate or delimit your area of study.

23

Ways in which you can do this:

- State what **interpretation of the key concepts** you are going to be working with (a kind of "working definition"). If, for example, your thesis will examine freedom of the press in reporting on murder statistics in the last five years, you might want to state that you will be working with a liberal notion of negative freedom, that is non-interference by government. Of course, your thesis might have as its aim to examine contesting interpretations. In that case, you delimit your study area to the different interpretations advanced in the literature. In the case of empirical studies, define your concepts clearly, preferably in operational terms (for example index calculations, types of observations) or in terms of persons, time and place.

- State what the **starting claims** of your thesis are. You don't have to reinvent the wheel; you can take certain claims as the basis from which your thesis will proceed. This is quite acceptable, provided you justify your starting assumptions in terms of previous research. For example, you may want to investigate the economic costs of treating pregnant mothers with HIV/Aids. You might state that your thesis will not engage with the debates about the causes of Aids, but will assume that HIV-positive mothers are potential Aids victims.

- **Narrow your field of study**. Most proposals suffer from being too ambitious and trying to cover too wide an area. It is through reading that you start to focus on particular aspects. Narrow down the study by limiting it to a particular group (for example women, or mothers, or single mothers, or single mothers with no employment, or single mothers with no employment living in rural areas), to a particular time (say, post-1994 South Africa), or a particular region (for example the Western Cape, the Cape Flats), or a particular kind of enterprise or occupation (say, spaza shop owners, informal caregivers), or a particular function (say, for example, generating staff morale as a function of HR; or instead of "media", focus on radio). Ask yourself: who, when, where, what, how, why? With each of the answers, you might be able to define your field of study more and more narrowly. These considerations contribute to the clarification of the unit of analysis that you are planning to work with.

This section is often merged with the section on the theoretical framework, since your starting assumptions or definitions are informed by the literature. However, some disciplines (especially in the natural sciences) prefer a separate section in which the limits of the study and definitions of key terms are spelt out. In many other studies, the very interpretations of key concepts are part of the discussion, part of the systematic investigation and clarification of certain meanings, and it would be inappropriate in such a case to "define" the key terms beforehand.

• •

2.10 RESEARCH DESIGN/RESEARCH METHODS

The research design section is often problematic in proposals. Many initial proposals tend to be too vague (not giving enough details about the actual ways in which the methods will be applied) and often are far too ambitious (covering too many methods or too big a field). So, when developing this section, be aware of pragmatic constraints (how much time, resources, and the like do you have available?) and be realistic in your planning (bearing in mind that you should research a contained area in detail, rather than a broad area superficially).

The function of this section is to give your reader a clear indication of the means by which you hope to achieve your research aims, to outline an appropriate procedure (which, if necessary, could be repeated by others), and to indicate the sources of data that will be used. The research design needs to go into some detail about the methods and procedures to be used. Up to now, you've been concentrating on *what* you are planning to research; here you need to consider *how* you are going to do so.

Many students think that, in order to do "proper" research, they *have* to undertake empirical data collection (do interviews, a survey, a questionnaire, an experiment, a case study, etc.). This is not the case. Rigorous research can also be purely theoretical – think of the many academic books and journal articles you have read that have not undertaken empirical studies, although they may report on empirical studies undertaken elsewhere. Your thesis can be purely theoretical or combine literature engagement with empirical work. This will vary according to the nature of your field of study, as well as the purpose of your research undertaking as articulated in your problem question/hypothesis formulation.

2.10.1 Purely theoretical theses

See section 6.1.2 for further details on this.

The majority of theses in Philosophy, Law, Language Studies, History and Religious Studies don't entail empirical research, but are theses that engage purely with theoretical or policy arguments and the development of theoretical insights. Of course, many theses in the natural sciences, social sciences, Policy Studies, Management Studies, Mathematics and Statistics are also purely theoretical and don't undertake empirical data collection. For example, a thesis in Sociology may compare different social theories on poverty or compare different research findings of studies written up in the literature. A thesis in Business Studies may explore the possible implications of a particular model of management within the South African policy of employment equity, or a thesis in Education may analyse the underlying assumptions of learner-centred education.

If you decide that your thesis will be a purely theoretical study, you need to do at least two things in this section of your proposal. You must specify:

* **what kind of sources** you are going to consult – policies, laws, curricula outlines, historical diaries, newspaper reports, minutes of meetings, "raw" statistics, original narratives, original reference texts, articles in journals, academic books, websites, written-up case studies, and so on.
* **how you are going to engage** with them – theoretical studies use research methods such as conceptual analysis, historical narrative construction, literary criticism, policy analysis, translation, critical literature review, semantic studies, discourse analysis, comparative literature or comparative statistics studies, meta-theoretical analysis, and so on.

25

- if necessary, the **feasibility** of obtaining the planned sources of information, that is, will you have access to your information/data sources? Some sources may be in the archives, or in restricted libraries – if so, state how you plan to gain access to these.

2.10.2 Theses that combine theoretical engagement with empirical work

If you plan to do empirical data collection, you must still have a theoretical (literature) base from which you proceed. In addition, you must do the following:

1. Specify **what kinds of method** you are going to use. Say whether you will use interviews, surveys, questionnaires, experiments, case studies, focus groups, observation, numerical measurement, or other methods.

2. **Justify your choice** of method. Tell your reader why it is an appropriate method to use for your specific purposes. Consider the pros and cons of alternative study designs, with particular attention to the control of extraneous factors that may produce bias.

3. Give details about the **where, when, and who** your method will involve. In other words, specify your sampling procedures – explain how you will determine the size, type and relevant importance in selecting the sample. Demonstrate to your reader that you know how to use the method properly by specifying the unit of analysis (for example, who or what are you going to sample?), how many will be involved, and in terms of what you plan to analyse your data. For example, if you are going to interview people, say why interviewing is appropriate for your research aims, what the purpose of your interview is, the structure of it (will it be structured, semi-structured?), as well as who, when and where you are going to interview.

4. Specify how you are planning to **verify** your findings. How are you planning to cope with non-responses or response errors? For example, if you are planning questionnaires, you cannot assume that people will necessarily be truthful or accurate in their responses. How will you test the reliability of the data you have collected? Will you use follow-up interviews, observation, triangulation?

See section 2.11.

5. Demonstrate, where necessary, that you are taking **ethical considerations** into account. When planning to involve human subjects or vertebrate animals in your empirical research, you must consult the ethical principles that govern research in your discipline.

6. Clarify the **feasibility** of your proposed undertaking. What arrangements have you made to obtain information and data? That is, will you have access to your information/data sources? Typically, students' research plans are far too ambitious (too many people to interview, too large a survey, too complicated a case study); show that you have thought realistically about what you are able to cope with in an academically rigorous way. Do you have the resources – equipment, funds, and so on – that will be required? Do you have the skills to use certain techniques and to analyse the data you will gather? If not, how will you learn

these? A word of advice: if you plan to use statistical analysis but don't have any experience in this, enrol for a course in statistics that will help you develop these skills so that you can apply them appropriately.

7. Clarify your plan for **data analysis**. For example, in terms of which criteria (informed by your literature review) are you going to analyse your findings? Make sure that you don't give the impression that you are going to gather the data and then think about the analysis afterwards!

A proposal for a traditional experiment will require detailed specification of the design, the variables, and the measurements that are going to be used. You need to pay careful attention to the details of the experiment so that it will give clear information with economical use of time, labour and resources. A good experiment is one that stands a good chance of delivering a clear result, decisive in the sense of proving or disproving the hypothesis. If you plan to undertake an experiment, ensure that your proposal addresses the following three main considerations when planning it:

- the costs involved (time, labour, resources)
- the ethical implications
- the theoretical and practical value of the hypothesis you're interested in.

There are styles of research, such as ethnographic, interpretive case studies, in which there is a principled resistance to pre-specifying the details of the research method. The design of the research is seen as *emerging* during the investigation. Proposals for this kind of research must demonstrate both the need for, and the right to, this kind of flexibility. The proposal must justify why the research questions are best dealt with in this way. You must also demonstrate, through argument and referencing, that you are competent to carry out this kind of research and capable of using the proposed methods (see Robson 1994).

In summary:

In this section you need to tell your reader about:

- your information and data sources (*where* you will get your information and data from)
- the methods and techniques you will use for obtaining information and data (*how* you will obtain the information and data)
- your justification for your choice of method/s and technique/s (*why* you will use, from the range of available methods/techniques, the particular ones you propose. Is the method appropriate for your research aims and questions?)
- where necessary, how you intend to verify the reliability of the information and data you collect
- ethical issues that may be involved and how you propose to address these
- the *feasibility* of your proposed undertaking
- your plan for *data analysis*.

Your research design must cohere with your aims, research question/hypothesis and research framework. (Ask yourself: Which design is most appropriate for the research question/s I aim to address?)

See Appendix 7 for research methods and sources.

Bear in mind that if you are writing *a minithesis*, it is a rigorous, systematic but *modest* piece of work. For Master's minitheses and Master's full theses you are not expected to formulate some *new* insights or develop some original knowledge. (This is a requirement only for Doctoral theses.)

Avoid being too ambitious in the number of methods you intend to use or the number of interviews/observations/experiments you are planning to undertake.

· ·

2.11 ETHICS STATEMENT

If your empirical research will involve people and/or vertebrate animals as research subjects, you will probably have to include an ethics statement in your proposal. Such a statement is an indication of your awareness of the ethical considerations and an agreement to conduct your research in accordance with ethical procedures. The goal of ethics in research is to ensure that no one is harmed as a result of your research activities.

Check with your supervisor about the ethical and professional guidelines for research in your particular study area. Also check with the disciplinary association (websites with the information are usually available). Most universities have an ethics committee which checks that the research proposes to follow ethical procedures.

Herewith are some ethical guidelines. The overall responsibility of the researcher is to

- design, conduct and report research in accordance with recognised standards of scientific competence and ethical research
- comply with national and provincial law and regulations and with professional standards governing the conduct of research
- minimise the possibility that the result will be misleading
- ensure there is no plagiarism; thesis authors must acknowledge scientific or professional contributions accurately

See Appendix 18 for useful websites.

- ensure that you have appropriate training and preparation for conducting the research
- obtain appropriate consent from the human subjects

- appropriately protect the rights and welfare of humans and the welfare of animals
- protect the identities and interests of those involved

- guarantee the confidentiality of the information given to the researcher
- consult the university Ethics Committee about any unclear ethical issues.

. . . .

2.12 CHAPTER OUTLINE

If the main research question is the heart of your research proposal, the chapter outline is the skeleton. It gives your reader an indication of how the various elements of your study fit together and the logical development of your investigation. It forces you to think about the way in which the sections are logically linked. Since your thesis is judged as a coherent whole, you must have a clear idea about the overall thread that binds the parts together. The way in which you structure your chapters depends on the kind of study you wish to undertake. The chapter outline is a kind of "contents page" which, in shorthand, reflects the entire development of your discussion. I always advise my students to spend time in getting this outline properly developed – it then provides a kind of structure for further readings and writing. If possible, note one or two key readings next to sections where appropriate. This will help you to get started on your writing. As noted earlier, the sheer volume of literature available can easily overwhelm you. It is tricky to judge what to include and what to exclude, or what to background and what to foreground. Having a clear and detailed outline – with listed key references attached to particular sections if possible – has the following advantages:

See
Appendix 3.1

- It provides you with a clear structure for writing and reading. You have a sense of where things should fit in.
- It provides you with a fruitful starting point for writing. You need such a map with a proposed destination.
- It provides your reader and supervisor with a synopsis of the intended development of your overall thesis.

Your thesis may be a study based entirely on the literature, or a study involving empirical work. Depending on what kind of study you plan to undertake, the chapter outline will differ.

2.12.1 Theoretical studies

In a theoretical study, each chapter addresses a particular aspect of the main research question. In order to formulate these aspects, go back to your title, keywords, aims and main research question. Using a mind-map or spider diagram, write the main research question in a box in the middle of the page. Draw a "leg" from this box to the top righthand side of the page, where you write an aspect that you will need to deal with in your thesis. Add more sub-legs, if necessary. Then, in clockwise fashion, draw another leg from the main box, with another important aspect that needs to be addressed. You may have to draw your spider diagram several times until you find a sequence and selection of aspects that works for your proposal.

29

Here is a simple example:

Figure 2.3 Spider diagram for research proposal

Source: Adapted from the NRF (2000: 43).

In Figure 2.3 each leg of the main box will be a chapter. Within the chapter will be various sub-headings as indicted in the smaller legs. The diagram is a way of helping you to identify key aspects and to structure the development of your thesis. For the proposal, write down the heading of each chapter (and some of the sub-headings within that chapter). There is no set number of chapters you need to have.

Another way of generating a chapter outline is to consider the keywords, aims and main research question. On a large sheet of paper, jot down as many sub-questions as you can think of – the kind of questions that you will have to address somewhere in your thesis in order to contribute to the overall response to your main problem question. You need to think long and hard about your research questions. Don't be reluctant to list questions that to you may be obvious. By putting them on paper you make them explicit and provide yourself with a checklist of the information and data that will be required to realise your research aim. Think of the research questions as the fuel and the motor that propel you towards your destination (research aim). Once you've done this, *group* questions that address the same issue, or that are logically linked. After you've done this, *sequence* the various groups of questions. Make sure that there is a logical development so that question 2 follows question 1, and so on. If you have about ten of these questions, approach each one as you would approach an assignment. List the various sources you will draw on in order to help you answer the question. If you write an informed 10–15 page "assignment" on each one of about ten listed questions, you will have your thesis!

Here is an example:

Main research question: **What are the moral justifications for Environmental Education (EE)?**

1. **What is the structure of moral arguments?**
   Why are moral frameworks necessary?
   How do moral frameworks inform self-understanding and practices?
   Is practical reasoning a form of moral argument?

2. **What are the current conceptual maps of EE?**
   What are the maps based on social theory?
   What are the maps based on moral theory?

3. **What constitutes a rationalist approach to EE?**
   What are its historical roots?
   How does disengaged reason shape self-understanding and practices?
   What is the justification for a rationalist approach to EE?

4. **What are the problems associated with a rationalist approach?**

5. **What constitutes a sentient approach to EE?**
   How did it develop out of a critique of the rationalist approach?
   How does the belief in the goodness of nature shape self-understanding and practices?
   What is the justification for a sentient approach to EE?
   How does a sentient approach claim to provide a better account of EE than a rationalist approach?

6. **What are the problems associated with a sentient approach?**

7. **What is a cooperative approach to EE?**
   How did it develop out of a critique of a sentient approach?
   How does social justice shape self-understanding and practices?
   How does a cooperative approach claim to provide a better account than a sentient and a rationalist approach?

8. **What are some of the problems associated with a cooperative approach?**

9. **Can the notions of attunement and practical wisdom improve a cooperative approach to EE?**

There are certain standard formats of thesis chapter outlines, but these are by no means mandatory. Although research needs to follow certain accepted conventions, it is also a creative process. Most research articles published in top international journals and leading scholarly books don't follow a rigid, set format, but demonstrate innovative structuring and ordering of sections. Consult these publications for inspiration and ideas on chapter outlines.

2.12.2 Studies that entail empirical research

The following generic structure was (and in some cases still is) used for theses that entail empirical research. Since many departments seem to insist on this formula, here is a skeleton outline of the headings and sub-headings.

See section 6.1.2 for further discussion of what sections each of the following chapters might include.

Chapter 1 – Overview and Rationale
1.1 Introduction and rationale or relevance of research
1.2 Problem statement and research question/s or hypotheses
1.3 Aims of the research
1.4 Theoretical perspective (describe and justify use)
1.5 Limitation of study and definition of terms
1.6 Overview of research design and method/s
1.7 Divisions of chapters

Chapter 2 – Literature Review
2.1 Introduction
2.2 ... etc: Grouping of literature under appropriate headings
2.3 Summary

Chapter 3 – Research Design
3.1 Introduction
3.2 Aims (or specific research problems and hypotheses)
3.3 Research method/s (description and justification of use)
  3.3.1 Ethical measures (if applicable)
  3.3.2 Measures to ensure validity and reliability (or trustworthiness)
  3.3.3 Data collection procedures
    • Sample (selection of subjects)
    • Method/s (if qualitative research, include a section on "the researcher as instrument")
  3.3.4 Data processing and analysis (if qualitative, include a section on "literature control"; if quantitative, include a section on "statistical methods")

Chapter 4 – Research Results
4.1 Introduction
4.2 Unit of analysis
4.3 Discussion of results
4.4 Summary

Chapter 5 – Conclusions, Recommendations and Limitations
5.1 Introduction
5.2 Conclusions
5.3 Recommendations
5.4 Limitations
5.5 Suggestions for further research
5.6 Summary

Many academics' response to the above structure is that it is a rigid, rather dull recipe that doesn't allow much space to demonstrate an innovative approach to research. Certainly, many contemporary empirical studies do not slavishly follow the above template. Lots of empirical studies opt for something much less "recipe" bound and are organised instead around pertinent problem questions relating to the research. Below is an example of a thesis that entails empirical research but in which the author shows that he has thought carefully about what each chapter will contain. The outline demonstrates creative control of the richness of possibilities in terms of which the research can be organised. Its rigorous and innovative structure of headings and sub-headings makes for much more interesting reading than the format above.

See also Appendix 15.2 for an example of a contents page of a thesis that entails empirical research.

Title: Internet Outreach Innovations in Community Health NGOs: a case study of the Mkambathi Nutrition and Primary Health Care Centre

Chapter 1: Introducing the study
1.1 Confronting technology and technological discourse
1.2 Clarifying the study's multi-disciplinary nature
1.3 Background to the study
1.4 Key assumptions guiding the research

Chapter 2: The Internet and computing: a search for illuminative approaches
2.1 Major historical trends in technology research
2.2 Critical perspectives in technology development
    2.2.1 Social Constructivism and social shaping of technology
    2.2.2 Information ecologies
    2.2.3 Actor Network Theory
2.3 Internet uptake and socio-economic development
2.4 Internet use in South Africa

Chapter 3: A case study of a Community Health NGO
3.1 The growth and context of Community Health NGOs in South Africa
3.2 A case study approach
3.3 The Internet Outreach project at the Mkambathi Nutrition and Primary Health Care Centre

Chapter 4: Technological innovation and tensions at the Mkambathi Nutrition and Primary Health Care Centre
4.1 Grappling with language and structural challenges
4.2 The influence of bureaucratic politics
4.3 Clarification of expectations and priorities
4.4 The road ahead

Chapter 5: Developing Internetworking for the support of Community Health care
5.1 Coping with national policy demands
5.2 Changes in the management structure of Community Health care centres
5.3 Communities of health care practitioners
5.4 Challenges of Internetworking in Community Health NGOs

Source: Adapted from Petersen, Y. (2002).

See section
6.1.2; also
Appendices 3
and 15.

> **In summary:**
> - Identify, for example through mind-mapping or in table form, the various sub-questions that the overall main research problem entails.
> - Group and sequence these sub-questions and sub-sections into chapters. Ensure that each of your chapters addresses an aspect of the main research question or hypothesis so that all the chapters taken together, in the end, provide an informed response to the main question your thesis addresses. (You can think of each chapter as an "assignment" with listed key sources. After having written all the "assignments", you'll have the first draft of your thesis in hand!)

2.13 TIME LINE

Here you need to outline a work schedule that couples the various research activities you will be involved in with a time frame. It is important that you present a realistic time frame that allocates sufficient time for the various activities and for revising, editing and producing the final text. Also, bear in mind that although you might be working in a linear way, some chapters will take much more time than others, and some sections, like the abstract, you'll write only at the end.

2.14 BUDGET

If you will be submitting your proposal to funders, such as the National Research Council (NRF) or Medical Research Council (MRC), you must include a budget. List what equipment or materials you need (computer, tape-recorder, scientific equipment, chemicals, etc.), as well as the kind of services you will have to pay for (transcription, photocopying, binding, postage, library loans, transport costs, etc.). Different funding bodies each have their own interpretation of what is permissible and fundable, and you will obviously have to adapt your proposal accordingly.

There is a general tendency to underestimate costs (and time), especially when money is scarce. Be as realistic as you can. Unforeseen circumstances will inevitably increase your costs and take more of your time than you have budgeted for.

2.15 DISSEMINATION OF RESEARCH

Some proposals have a section called "Dissemination of research". Check with your supervisor or the chairperson of the relevant committee whether you need to include this. It is usually not necessary for a thesis at Master's level.

34

Address the following questions in this section (Conradie 2000: 12):

- What will you do with the results of the research once the project is completed?
- How will you make it available to those who may benefit from the research?
- Do you plan to publish extracts from the research in accredited journals, or in popular journals?

Avoid making grandiose claims about your publication and dissemination plans.

• •

2.16 PRELIMINARY BIBLIOGRAPHY

List all the texts that you have referred to in your proposal, as well as others that you have consulted thus far. The bibliography need not be extensive at this stage, but it should provide an indication of the texts that are important and relevant to your project.

There are many different referencing conventions. Some disciplines adopt specific conventions, so check with your supervisor which convention is acceptable in your field of study. Or you can consult the main journals in your proposed area of study and use those referencing conventions as your template. However, one golden rule is that you must remain *consistent*. Don't mix different conventions – this is especially hazardous when you have written down various references from different sources with different conventions. Adapt all entries to *one* convention.

Whichever convention you decide to use, you must ensure that your bibliography is technically faultless. Check every entry carefully. Proposals with faulty bibliographies are sent back by the committee. The committee judges that if you cannot manage to do a bibliography correctly (after all, this is a simple technical exercise), then you will not be able to manage the more sophisticated academic skills needed for research! Sloppiness in technical tasks is inexcusable.

See Appendices 3.1 and 3.2 for examples of bibliographic conventions, and Appendix 18 for useful websites on referencing conventions.

In summary:
- Find out which referencing convention is used in your intended research discipline.
- Be consistent and ensure that all entries conform to the adopted convention.
- Since it is a preliminary bibliography, you need list only the key texts at this stage.
- It must be technically faultless.

Part B

Negotiating and
sustaining the supervision
and thesis writing process

Now that you have successfully submitted your research proposal, you need to focus on writing the thesis itself, under the guidance of your supervisor. Part B is a collection of pointers that may assist you to develop a productive academic relationship with your supervisor. This part does not try to prescribe ways in which supervisors ought to act. It does, however, try to provide some advice as to what you can realistically expect from your supervisor and what your supervisor can expect from you. The supervision process differs across various disciplines, so the guidelines offered here are wide-ranging and in many cases may not be appropriate for your subject-specific purposes.

There is, however, one central idea that underlies this part: it is the idea that successful, pleasurable supervision is based on both student and supervisor clearly and explicitly understanding what the relevant processes and issues are. Part B tries to help you in this. It draws on "good practice" developed by various supervisors in different disciplinary areas, and hopes to give some pragmatic advice on how to structure the process in a productive way. A successful supervisor can maximise the student's chances of completing a thesis quickly and efficiently, but a student also carries great responsibility for seeing it through. Much of Part B draws on the book by Delamont, Atkinson and Parry (1997). This is well worth a read, especially for Doctoral students. The thrust of this book is that higher-degree work should not be based on ad hoc criteria, "approached casually in the interstices of the working week. It demands and deserves to be treated seriously as a set of commitments and demands on a par with other teaching and scholarly activity." (1997: 13)

Part B is organised so that it follows the stages of the supervision process. Of course, these stages do not necessarily proceed in a linear fashion, but they work for the structure of this section of the guide. Few supervision processes happen without a hitch or without disagreements. Working relationships develop over time, in a non-linear fashion, and cover many facets of academic work. Part B offers a starting point for developing intellectual pleasure from the supervised writing process.

3 Getting off to a good start

3.1 THE MOVE TO MASTER'S AND DOCTORAL STUDIES

For many the new status of a "Master's" or "Doctoral" student can be unsettling. It may seem as though everyone else knows what to do and what is expected and you may feel too shy to ask in case you show your ignorance. The growth in structured Master's degrees, made up in part by coursework, can clarify what constitutes "academic work" at this level, but there is still a lot of uncertainty about the expectations of a thesis or even minithesis. The requirement of a Doctoral thesis, to be "original" and to "make a contribution to knowledge", can be scary and perplexing. As an undergraduate, you have probably learnt fairly standard ways of reading and writing or following a set of standard experiments designed to work. The transition from mainly discursive writing to critical commentary or engaging in original theorising is a major one. Similarly, the move from doing laboratory work with experiments that are designed to work (that's why they are chosen as part of the curriculum) to designing and conducting experiments that encounter novel experiences, is one that can make you feel insecure and uncertain of how to proceed (or even wonder whether it is worth proceeding at all).

Maybe you are working with a novice supervisor who might also feel uncertain about what is expected. Understanding the supervision process and knowing how to make judgments about the academic worth of a thesis are skills that take time to develop. There is no substitute for experience, but there are some ways in which you, as a new postgraduate, can acquaint yourself with particular requirements and, in so doing, reduce uncertainty.

As a new postgraduate student, you should

- locate available sources of advice and support (for example Master's or Doctoral students who have completed their theses, other postgraduates and supervisors working in the research area, university guidelines, books on how to write a thesis, Internet sites, training sessions in specific methodologies, and summer school courses)
- borrow theses in your disciplinary field from the university library and see what a "successful" thesis entails
- borrow theses from other universities and develop a sense of the shared academic standards that govern thesis work

- consult books on how to complete a PhD or on how to succeed in your Master's studies (there are numerous copies of such guides in most university libraries)
- consult your university's guidelines for examiners of theses (knowing the criteria in terms of which your examiners are asked to judge your thesis will help you structure your thesis accordingly; copies of these criteria are usually available from the Exams office or the relevant university committee that administers examination of theses)
- ask to work with students who are reaching the final stages of their theses; this will provide you with both guidance and motivation.

See Appendix 8 for some criteria for judging a thesis.

3.2 TEAMING UP WITH THE BEST SUPERVISOR

The relationship between student and supervisor is one in which both parties invest time and energy. It is therefore important to select a supervisor with whom you can work productively. Supervisors, in turn, choose students by their marks, their writing, their originality, their proven work capacity and determination – in a word, ability. And students should do likewise. You should be encouraged to choose the toughest, brightest supervisor in your field. Some departments are steering towards supervision committees, that is, groups of academics who collectively take responsibility for direction, support and feedback. However, most departments operate on an individual supervisor-student basis. In the case of a Master's minithesis, co-supervisors are not usually appointed, but for a Master's full thesis or a PhD, co-supervisors are appointed where necessary. The matching of student and supervisor is usually based on the following considerations:

- Which student does the supervisor want to work with?
- Which supervisor does the student want to work with?
- Which supervisor is an expert in the field that the student wants to work in?
- Which supervisor is the toughest, brightest supervisor in that field?
- Which supervisor is available (not on sabbatical)?
- Which supervisor is willing (that is, has the space and time to take on a new student)?
- Is there is an existing research project that the student can slot into (many supervisors employ postgraduate students to assist with research on a larger project)?
- Does the proposed thesis that the student wants to work on fit in with other theses that the supervisor is supervising?

3.3 DETERMINING WHAT IS TO BE DONE

Many of the problems in the supervision process can be avoided by clearly setting out the expectations that both parties have of the relationship. This should be done right at the start and, as your needs as a student change over time, the agreed rules of the relationship need to be discussed periodically.

At first glance the supervision process seems to entail just you and your supervisor, but many more parties are involved, albeit not as overtly. The process also involves the university and the broader disciplinary community. Supervisors and the university are in an accountable relationship – what does the university expect from its supervisors? What are the department's expectations? In turn, what support can you and your supervisor expect from the department and the university? On a broader spectrum, you and your supervisor are also accountable to the discipline. What does the discipline expect from you? What criteria does the discipline have for academic work?

At the first session you have with your supervisor, ask to discuss how the two of you will work, separately and together. Explain how you like to work and see whether that will fit in with your supervisor's schedule. The following need to be agreed on, (see also Delamont et al. 1997: 17–23):

3.3.1 Practical arrangements

The best time of day to meet

Decide on a time that fits into both your workloads. If you are a part-time student, you can usually meet only after a full day's work, which is not always good for clear, focused thinking. Explore the possibilities of being able to take time off from work to meet with your supervisor or of meeting over weekends every now and then.

Where to meet

Some places might be more conducive for a productive meeting than others. Agree on a place with which you are both comfortable, be it in another office or suitable venue off-campus.

How often to meet

It is important to schedule *regular* meetings, but their frequency might vary depending on the stage of the thesis. Initially it might be fruitful to schedule weekly meetings, even if they are short, because this is the stage where it is easy for you to drift off course during the beginning phases of the thesis. Thereafter, a meeting every second or third week should be enough. As a guideline, about 30 supervision meetings a year is a sensible target (Delamont et al., 1997: 18). Set up all these meetings at the beginning of the process, so that you can both plan "around" these dates, rather than trying to fit them in at a later stage. Be realistic and bear in mind the busy times of both you and your supervisor (for example exam times for the supervisor or job-related peak times for you), or when there are other commitments (for example the due date of a baby, a planned conference trip or holiday). Regular meetings serve to keep the momentum. They signal that the process has stages and deadlines by which to measure progress, or the lack of it. Regular meetings also assist to identify problem areas early on, before these become major obstacles to the completion of the thesis.

Setting an agenda

At the end of each meeting it is important that a clear agenda for the next meeting be set. Both of you should contribute to the agenda. For example, the supervisor may at one time suggest, "Next week, please bring X and we'll discuss it". At other times you need to set the agenda and suggest what you would like to focus on next time you meet.

The mechanics of cancellation

One of the most maddening things is for one of the parties not to show up for the meeting. Make sure that you are both clear on the mechanics of cancelling a meeting (telephone each other at home? What number? How long in advance?) for those times when cancellation is unavoidable. Reschedule immediately.

Keeping a record

Keep a record of the dates of your meetings, the key issues discussed and the agreed tasks for the next meeting. Keeping a record signals that you have high expectations of the process and it gives you a sense of direction. In cases where other supervisors have to take over, an objective record will be valuable.

3.3.2 Reasonable and mutual expectations

It is important that your supervisor is explicit about what he or she is able to provide for you, such as advice on the literature search, the selection of method, practical help, assistance with securing funding, the loan of equipment such as a tape-recorder, gaining permission to access certain sites, honest and timely feedback, tea and sympathy. At the same time you need to be clear about what he or she cannot provide for you, for instance language editing (correction of grammar and spelling), advice on how to use certain software (such as SPSS), literature advice on a certain aspect of the topic, and so on. If you, for example, need help with statistical analysis which your supervisor cannot provide, arrange to meet with someone who can assist, or enrol for a statistics course. It is important that you take the initiative for many aspects of your thesis and your own academic development. Relying on your supervisor to provide it all will lead to frustration and a probable breakdown in the supervision relationship. It is especially important for you and your supervisor to set clear expectations about your writing, the form in which it should be and the times when written work needs to be submitted. In turn, ask your supervisor about when you can expect feedback and the form it will take. Realise in advance that the harsh feedback your supervisor might give you is not a personal affront. It is not, for instance, the supervisor's job to correct your grammar, punctuation, spelling or referencing notations – these are technical requirements for which you must take responsibility. The primary task of supervisors is to ensure the academic rigour of your research, and so all copies of writing submitted to the supervisor should already have been proofread and edited for technical correctness. If you need assistance with this, you can do the following:

- Find out the sources and support systems available that can help you. Most universities have centres that assist students with academic writing. Make a series of appointments.
- Employ a professional proofreader to work *with* you in editing and correcting the work.
- If you are writing your thesis in a language in which you are not fluent, enrol for a language course.
- When generating your text with a computer keyboard, use the spell check and grammar check (this will not give you complete correctness, but will at least eliminate many unnecessary errors).

The following are some reasonable expectations, taken from Delamont et al. (1997: 24):

Reasonable student expectations

A student can expect the supervisor to

- give regular supervision
- give timeous written feedback
- be reasonably available and approachable
- respect meeting times (to defer interruptions such as the telephone and visits)
- be constructively critical in the feedback
- give academic guidance on structure, content and reading
- have a good knowledge of the research area and to direct you to sources in areas where he or she does not feel qualified to advise
- deliver on the mutually agreed and scheduled tasks.

Reasonable supervisor expectations

In turn, a supervisor can expect students to:

- turn up for appointments, prepared for them
- write regularly, and share the draft material
- tell the truth about work done and not done
- deliver texts that have been proofread for spelling and grammar errors
- keep in touch – practically (inform of holidays, sickness, change of address) and academically
- do the research tasks that have been mutually agreed upon and scheduled.

See Appendix 9 for an example of a contract entered into between student and supervisor.

3.4 CO-SUPERVISORS AND OUTSIDE SUPERVISORS

Co-supervisors are usually appointed when the content and/or research method call for specific expertise that warrants the appointment of a specialist in that area. In cases where students are located at a distance from the university, co-supervisors located in institutions closer to the student may be appointed. So, if you are in need of specific expertise that your supervisor cannot adequately supply, or if you are a distance student, you may request that the university appoints a co-supervisor.

In order to prevent your supervisor and co-supervisor from giving conflicting feedback, it is a good idea for a list of division of tasks to be drawn up. You, your supervisor and co-supervisor must be clear about who will be responsible for what. In cases where your supervisor and co-supervisor disagree, a discussion needs to take place and a decision made about how you should progress. Supervisors and co-supervisors should maintain frequent contact; ask that they at least copy each other on all email with you. In this way, the risk of conflicting advice is minimised.

It happens sometimes that, halfway through your thesis, your supervisor may leave for another job or go on sabbatical. If your supervisor goes on sabbatical, try to arrange for contact via email, with a clear work agenda drawn up before departure. If, however, your supervisor leaves to take up a position elsewhere and is unable to see through the completion of your thesis, a new supervisor may need to be appointed. This increases the risk of different advice and a different thrust to the development of the thesis. Depending on how far you have progressed, a new supervisor may suggest radical changes. In such cases, it is imperative that open and honest communication be maintained.

As a postgraduate student, you should keep a record of your work agenda with your supervisor in case he or she should go on sabbatical or a new supervisor should be appointed.

• •

3.5 BEING SUPERVISED FROM A DISTANCE

Being a distance student presents a number of challenges: maintaining frequent contact with your supervisor in order to monitor your own progress; interacting with other students and academics for discussion; networking with others to keep up your own motivation; and coping with a sense of isolation. There are a number of strategies you can adopt in meeting these challenges.

Some universities and departments have many students who are located outside the region of the institution (some even in other countries). In some cases, where a number of students are grouped in a geographical region, a support network has been established. These support networks might consist of informal groups that meet regularly to present work-in-progress for discussion with fellow students, or it might be a more formal structure with appointed mentors, paid by the university. If there is such a mentor network within your region, join it and use the space for intellectual interaction. If there is not such a network within your region, contact the university to find out if there are sufficient students in your region to warrant the establishment of such a network. If there are not enough students to set up such a formal arrangement, or if the university is not prepared to appoint a mentor, then ask for the university to give you a list of names and contact details of students within the area. Then, on your own initiative, set up a regular meeting group. I have seen such groups operating very productively, even if students are from different disciplines and are a mixture of Honours, Master's and Doctoral students. It is an intellectual forum in which you need to articulate your ideas and present your work. That exercise in itself is immensely valuable – you only really get clarity on your own thoughts when you have to communicate them to others. If there are not enough students to set up such a group, and if you are in a region near another tertiary institution, ask your supervisor to contact a staff member of that institution whom you could approach for support.

Isolation is a key reason for students not progressing on their thesis. It is therefore important that there should be regular communication between you and your supervisor, or between you and other members of the academic community. The most fruitful medium of communication is email. This is speedy. Join an academic on-line chatgroup within your discipline. Ask the department to set up a list server for the group for regular email discussion. In cases where email is not available, telephone, fax, postal and courier services can be used. You must keep copies of all posted material; too often packages get lost.

Students located elsewhere should apply for permission to use libraries of other institutions. You will usually need a letter of permission from your home university, stating that you are a registered student. Sometimes a small joining fee is required (for a library card).

3.6 SETTING UP STRUCTURES

The research proposal is a plan of the research work to be done and of the writing of the eventual text. It is an indication of *what* you will investigate, *why* it is an interesting issue, *how* it will be investigated, and *when* various sections are going to be completed. Of course, few final theses mirror exactly the initial research proposal, since research is substantially investigatory work during which certain things become clear or are discovered. There are often unexpected revelations. Periodically you and your supervisor might need to re-look at the "thread" of the thesis and decide how best to incorporate the new finding, or insight, and to adjust parts of the thesis accordingly. No one compares the final product of the thesis with the initial research proposal – the thesis is judged on its own merits. The research proposal is a necessary tool to get started and to direct the process of investigation, as well as to remind you of the overall layout of the thesis and its main aim.

In starting your thesis, constantly bear two golden rules in mind (Delamont et al. 1997: 121):

> **1. Write early and write often because:**
> - the more you write, the easier it becomes
> - if you write every day, it becomes a habit
> - tiny bits of writing add up to a lot of writing
> - the longer you leave it unwritten, the worse the task becomes.

45

2. Don't first try to get it right; get it written because:
- until it is on paper, no one can help you get it right
- you find out what you think through writing
- drafting is a vital stage in clarifying thought
- as you draft sections, other bits become clearer
- drafting reveals the places where it isn't right yet, in ways that nothing else does.

Every student and supervisor knows of some horror story of drafts that got lost, stolen, burnt or destroyed. So work on a safeguard principle from the beginning. Make back-up copies of your disks (clearly labelled and dated so that you know which version is the most recent one) and paper copies of your written work (again, clearly dated). Data needs to be backed up professionally. It is a prudent idea to keep these back-up copies in another place from the originals – at home, or in the office or with a friend who will store them safely.

Determine from the start if there are technical skills that you need to develop. If you are expected to write on a computer (almost all institutions insist on this now), but feel uncertain about the medium, enrol for a basic word-processing course. If you need to use unfamiliar software (for example SPSS or Research Toolbox), you must develop those skills sooner rather than later. Similarly, if you need to become more fluent in the language in which you will be writing your thesis, seek advice or enrol for a course as soon as possible. Set a time line for when you plan these skills to be developed. Ensure that these skills are appropriate and adequate for the thesis. If you need to know only word processing in order to complete the thesis, don't overextend yourself by acquiring sophisticated skills in graphics as well. It's easy to become distracted from the main task, that is, getting the thesis done.

Organise the practical matters of the project – do you need access to a computer, space to work, specific equipment? Set up a schedule of bookings with the computer lab if necessary. If you are working at home, set up a dedicated space that is yours – working on the dining room table is not advisable since you will have to pack up everything periodically and that prospect can tempt you not to continue working on your thesis! When I was writing my thesis, I had wall space onto which I could stick pieces of paper with the main research question of my thesis (in bold letters!) and the outline of each chapter on which I could make notes as I progressed. In this way I was constantly reminded of the thread that ran through the thesis, giving my writing direction and helping me to maintain focus. Arrange with your family to protect particular times (and space) when you will be working on your thesis. Set up the routine right from the start, instead of trying to fit in thesis work on an ad hoc basis in between family time. If you can't work at home, find a suitable work space on campus, or in the public library, or – as some of my colleagues have done – in a quiet coffee shop!

3

3.7 GETTING GOING

In the early stages, many students flounder. The project can seem so daunting that you often don't know where or how to start. Here you need to ask your supervisor to give you some definite tasks, guidelines and activities. Ask your supervisor to give you a contained list of readings to start you off (which are readily available in the library). Then review each one in written form. In the beginning, agree on a certain page length or word count for each review. It helps to set preliminary boundaries at a time when everything seems loose and open. Moreover, completing these small tasks gives you a sense of progress and helps to build confidence.

Often students have developed a good research proposal and then don't quite know how to "operationalise" it. This can be due to simple things such as, "Where do I go? What do I look at?" Find out what support services and structures are available on campus and ask your supervisor for pragmatic suggestions to assist in locating these. If necessary, ask your supervisor to help you write letters to access certain sites, set up meetings with relevant people, and so on. As a postgraduate student, you are expected to take far greater initiative in accessing information and developing your own studies, but that doesn't mean you have to do this all by yourself. Your supervisor is there to guide and assist you (but not to do it for you!).

• •

3.8 WHAT KIND OF STUDENT ARE YOU?

Even the most able of students need reassurance about the quality of their work and the development of the process. Without reassurance, it is difficult to progress. Delamont et al. (1997: 31) note that there are three predominant kinds of students. They call them the cue-seekers, the cue-conscious and the cue-deaf. Try to identify, honestly, which kind of student you are. Doing so will alert you to the kinds of support you need and to ways in which you might need to change in order to become more of a cue-seeker and so increase your chances of success.

Are you:

1. **a cue-seeker** – the fewest in number? Do you deliberately interact with the system, actively seek out information about thesis requirements, engage staff in discussion, and set out to impress lecturers with your ability and level of interest? Are you self-motivated, self-disciplined and, to a large extent, self-directed? If so, your supervisor plays an important, albeit less interventionist, role.

2. **cue-conscious** – alert to hints about assessment, understand the requirements, but don't actively seek to acquire additional knowledge to that which the supervisor gives? Do you deliver what is expected of you but need motivation, direction and a clear agenda of how to proceed and what to do? If so, your supervisor needs to keep a direct handle on the progress and development of the various stages in the thesis production. Develop more initiative in the development and progress of your own studies.

3. **cue-deaf?** Do you often not "hear" what your supervisor is saying and often

47

want to "revise everything" because you have not picked up on the supervisor's guidance and selectivity? You may be hard working, but your work is not always efficiently focused because of your lack of understanding of the advice or rules pertaining to the thesis. Cue-deafness in foreign students might also be a result of difficulty with English or because of lack of familiarity with social conventions. Criticism from the supervisor does not necessarily mean that the work is entirely useless or that the criticism should be taken personally. If you are cue-deaf, ask your supervisor to spell out aspects of the supervision process or rules and agendas rather more explicitly than might otherwise be the case. Try to become more of a cue-seeker by taking the initiative.

• •

3.9 SUPERVISOR INTERVENTION AND STUDENT AUTONOMY

The level and extent of intervention required is dependent not only on what kind of student you are, but also on what kind of thesis you are writing. Master's minitheses are conceived of as modest pieces of work, limited in scope and depth, with the same academically disciplined and systematic writing as a full thesis. Master's full theses are more substantial in length, scope and depth of discussion. Typically, supervisors direct the Master's thesis development, advising students quite clearly which steps to follow. A Doctoral thesis, however, is expected to make an original contribution to knowledge and, as a Doctoral student, you are expected to take a lot more responsibility for the development of the argument, the seeking out of information and the pace of the thesis production.

Negotiating and maintaining the balance between "heavy-handed dominance" by your supervisor and a "hands-off" neglect is a delicate matter. Different supervisors have different styles, ranging across the spectrum. Questions that students generally ask are: "How much can I expect my supervisor to guide me?"; "If the supervisor doesn't correct my language, does that mean it is of an acceptable standard?"; "What should I do if I disagree with my supervisor's comments or suggestions?"; "To what extent should I steer the thesis development and to what extent can I rely on my supervisor to ensure that it is academically sound?" Questions that supervisors generally ask are: "How much should I be spoonfeeding?"; "What is the line between advising students of certain formulations and writing the thesis for them?"; "To what extent do I find suitable literature for the student?"; "What kind of interventions silence and demotivate the student?"; "Should this student aim merely to pass or go for a distinction? If so, to what extent should I intervene?"; "Given the push for completion and throughput rates, to what extent should I control the thesis production?"

These are tricky questions with no clear answers because the supervision process is a complex and difficult one. There are expectations of you the student, your supervisor, the university, the funder of the research, and the academic community of the discipline in which the thesis is being written. Juggling all these considerations is too complex a task to capture in a few "tips", but there are some useful guidelines:

GETTING OFF TO A GOOD START   CHAPTER

3

- In the early stages of supervision, agree on what your responsibilities are and on what are those of the supervisor, bearing in mind the kind of student you are, what kind of supervisor you have, the kind of thesis being written, the academic demands of the discipline, and the constraints put up by funders.
- If necessary, in the early stages ask your supervisor to set clear tasks and guidelines. Ask to be directed in your reading through the recommendation of some key texts. If needed, ask for help to break the huge project into bite-size bits. Completing small, short tasks engenders a sense of progress, enthusiasm and confidence.
- If you are part of a funded research project with definite deadlines, agree on the tasks you need to do and the kinds of support you will be given to meet these deadlines. Discuss how and to what extent the thesis work must cohere with the project demands or with the work of the others on the project.
- You can ask for extensive assistance in the early stages of the thesis writing, but you will be expected to develop self-reliance soon thereafter.

(As mentioned in the preface, I often compare learning to write a thesis to learning how to ride a bicycle. You can listen to other people telling you how to do it; you can watch how others do it; you can learn from others about the various parts of the bicycle and their functions; and you can read "tips" on how to ride a bicycle. But ultimately the only way you really learn to ride is to get on the bicycle, with a steady supporting hand holding the saddle, and to wobble along, at times crashing and getting back on, and proceeding each time with a little more confidence until the steady hand on the saddle lets go and you're on your own! No one can ride the bicycle for you; you must do it yourself, and the more you do it, the better you become.)

In summary:
- Draw up and sign a written contract with your supervisor (and co-supervisor), which clearly spells out arrangements for regular communication. Communication can be via email, telephone, fax, or post.
- Get onto email.
- Join an academic on-line discussion group in your field of study.
- Encourage your home department to draw up a list server for regular email discussions.
- Join or establish a support network if you live in an area where a number of students are grouped.
- If you're a distance student, arrange a site visit from your supervisor if possible. An agenda about what preparation you should do will help make the contact meeting fruitful and focused.
- Make arrangements with an academic from another institution near you to assist with intellectual support.
- Arrange to gain access to libraries elsewhere if you don't have easy access to the home library.

© Van Schaik
Publishers

- Revisit the layout and sections of the thesis from time to time with your supervisor to remind yourselves of the trajectory and main aim of the thesis.
- Get into the habit of making back-up copies of work and storing them in safe places.
- Identify what skills need to be developed, what gaps need to be filled, and set a time line for acquiring these.
- Set up a suitable physical workstation with the necessary equipment.
- With your supervisor, decide on definite, small tasks.
- Draw up a "start-up" reading list.
- Make some pragmatic arrangements.
- Find out what available support services and structures you can draw on.

3.10 THE DIFFERENCE BETWEEN BEING A MASTER'S AND A DOCTORAL STUDENT

A substantive difference between a Master's and a Doctoral thesis is that the work done at Doctoral level should show proof of "original work". This means that, as a Doctoral student, you must demonstrate independence of thinking and writing. Whereas at Master's level the supervisor may be fairly interventionist with regard to thesis topic, research design, and development of argument, at Doctoral level you are expected to take much more responsibility for this yourself. At Doctoral level, the supervisor is a critical discussion partner, rather than a director of thesis production.

Working with focus

If you were to talk to students who have completed their theses, many will admit that it is all too easy to drift and allow valuable time to pass unproductively. If you start to feel aimless, or feel that your work is not progressing, then your self-confidence and enthusiasm can wane seriously. If, however , you make a "flying start", your confidence and enthusiasm will grow.

4.1 THE LITERATURE SEARCH

Whatever the discipline, you have to come to grips with the literature – this means learning how to find it, read it, assimilate it, make informed judgments about it and write about it. The literature search remains a headache for many. The infinite possibilities of extending the thesis in various ways mean that it is difficult to know when to stop. Often the most difficult decision about planning your research is not the decision about what to include, but rather what to *exclude*. So the question about what a reasonable literature search is depends on the topic and the problem. (I once supervised a minithesis that had only eight references in the bibliography – the student did an in-depth analysis of a particular philosopher's concept of ethnocentricity.) You need to keep a balance between reading widely and maintaining focus by keeping to the central purpose of the thesis.

4.1.1 How to get started

Some theses have a literature review chapter, even if it is called something else; other theses develop an argument and draw on appropriate literature throughout the thesis (these are usually analytical, philosophical, conceptual or literary studies). When there is a chapter that reviews pertinent literature, this usually comes early on in the thesis. You should ask your supervisor in the initial stages of your literature review to assist you in accessing fruitful sites and finding relevant literature. As the thesis progresses, you will start to develop familiarity with certain sites (libraries, journals, online sites) and will need less intervention from your supervisor about what to read or where to find it. If you are part of a larger research project or part of a discussion group, ask your fellow students to assist you, especially in the beginning, with accessing sites and finding relevant literature.

See Appendix 3.1 – "Chapter outline" section, and Appendix 15.

51

 Van Schaik Publishers

One of the first things you should do is to set up a system that will keep records of all the texts consulted. Most thesis students can tell stories of quotes with "lost references" and spending days trying to track down details of a specific article or book. Especially when you are working towards a deadline for submission, the last thing you want is to be held up by searching frantically for details of references that you consulted ages ago. Keeping full, clear records from the start can help to eliminate such occurrences. I often advise my students to keep a stack of blank (8 x 12 cm or larger) library cards with them and, whenever they have found a useful text, to write down the full details on the card, including where they found it. If they should have to refer to it again at a later stage, at least they will know where to go. Alternatively, you can use various software packages such as Research Toolbox or Endnotes to help organise your references. The advantage of using such a software package is that it automatically formats and alphabetises your references for you. The disadvantage is that, should you lose the file (through a virus or a stolen computer), you may have lost everything. Keep hard copies. Also, when you are working in the library or in a place where you don't have a laptop available, you still need a system whereby you can keep a record of the references consulted – either on cards or in a notebook.

Librarians can be extremely helpful in finding suitable literature. Meet with the specialist librarian who looks after your specific discipline's library needs. Liaise with your specialist librarian about interlibrary loans and other services, the library's on-line search engines, and computer databases of research. Librarians can show you how to use these effectively, where to find lists of abstracts (such as those in the ERIC retrieval system) and where past theses are kept. Consulting successful theses is useful for finding out the kind of scope and depth required for a thesis, as well as for "piggy-backing" on the bibliography of a thesis on a similar topic. Moreover, once a good working relationship is established with the librarian, you can ask to be informed about new books or journal articles that have arrived that might be relevant to your topic.

Occasionally you might discover that you cannot find any literature on your topic. This may be because your university's library is understocked, or because you lack vision and are defining the topic too narrowly. In the case of inadequate library stocks, first of all ascertain from the specialist librarian that this is in fact the case. Many libraries stock journals and books in electronic form, so they only *look* understocked. If the reference you are looking for is indeed not available in your library, ask the librarian to steer you to other libraries, interlibrary loans and to the Internet. As universities all over are coping with cuts in funding, many institutions situated in a common geographical region (such as the Western Cape or central Gauteng) have agreed to share their library resources. So, if you can't find adequate material in your own library, the chances are that you can access the library stocks of neighbouring universities. In the case of a too narrowly confined topic, you might have to read *against* and around the *topic*. This can often be a creative undertaking in bringing to light some interesting aspects. Or you might have to rethink the topic and perhaps foreground different things on which there is more literature. Of course, a genuine lack of literature on the topic might be a creative opportunity for the thesis to highlight gaps in the research field, to give information on biases, preoccupations and blind spots in the research community and to try to offer an explanation for these. The thesis itself can become a principled criticism of the state of the field.

4.1.2 Requirements of a literature review

Once you have located appropriate literature, you need to read and make notes. Of course, you already have experience in writing assignments in the discipline and are familiar with the kinds of academic practice and conventions that characterise your field. As a research student, especially at Doctoral level, you are first and foremost a reader who needs to know *how* to read and *how* to write coherent accounts of what you have read. The literature review is a demonstration of your ability to do so. In order to develop this academic skill, you need to know what adequate note-taking involves, from analytical summaries to short précis.

For theses that have a literature review chapter, the literature review has the following requirements:

- It should show the reader that you are capable of searching for relevant material, summarising it, arranging it by various themes and relating it to your own work.
- In the case of a Doctoral study, it should show that the thesis is original or is a principled, conscious replication of a previous one.

See section 5.1 which deals with what constitutes a critical discussion.

Note the following:

- The literature review should not leave out important key texts.
- If appropriate, the review should include current texts.
- Readings should be grouped together for a clear purpose.
- The review should avoid being dull.

In order to write a review that fulfils the above requirements, you should be in constant dialogue with your supervisor, other students and with the specialist librarian. Scan bibliographies of published authors and the contents pages of new editions of the key journals in the field. In order to avoid being boring, reviews should not just be long lists or sequences, but should be arranged by theme, and form a coherent *base* for the particular issue the thesis is investigating. Most important, you should demonstrate that you know how to make sense of the literature, not just to list it. The thesis must show that it *uses* the literature to highlight and inform the specific issues it addresses – it must show that it can derive and explore ideas through the literature rather than merely passively report on it.

Demonstrating familiarity with referencing and citation conventions is a crucial aspect of the thesis. There are many different conventions, some discipline-specific. Ensure that you adopt an appropriate one for the discipline and then be *consistent* in its use. In order to find out which convention is appropriate for your discipline, refer to the key journals in your field and adopt that convention. Either photocopy a page of references as a template you can follow, or refer to the "Instructions to "Authors", usually on the back inside cover of the journal, where the referencing conventions are stipulated. Alternatively, you can rely on a

© Van Schaik Publishers

suitable software package that will format your references automatically! This is handy and timesaving (although some academics find that doing it this way never trains students in the discipline of constructing a bibliography themselves).

As a student you must know what constitutes plagiarism and how to avoid it. One quick way to get your thesis turned down is to be academically dishonest through plagiarism. Ask your supervisor to help clarify the balance between "your own informed voice" and plagiarism. Be clear about the notion of intellectual ownership and ensure that substantive claims and contested premises are properly acknowledged. Just to what extent one must acknowledge is often a discipline-specific issue – for instance, the fact that Van Riebeeck landed in the Cape in 1652 is now part of "common knowledge" and, as such, need not be substantiated with a reference, but the "fact" that apartheid South Africa had a capitalist economy (rather than a centralised command economy) is clearly a contested claim and, as such, would have to be substantiated or shown to be false.

Unless your thesis addresses a problem for which you don't need current literature, ensure that your discussion of the literature includes up-to-date sources. Demonstrate that you know what the current academic debates on your topic are about.

4.1.3 Developing your first draft of the literature review – what goes where?

One of the major difficulties that many students run into when they are developing their literature review is to decide on what goes where. Does a literature review chapter mean that you first have to summarise all the relevant literature up front and then proceed with your "own" discussion? Or do you "critically engage" with each text as you refer to it? And where do you fit in the literature review about the method you are planning to use? After all, you must not only show *how* the method has been used in previous research (written up in the literature), you also need to argue *why* this is an appropriate method for your purposes. So does the review of the method as written up in the literature come into the literature review chapter or into the methodology chapter? And if you have opted to work from the "outside in" (that is, apply a grand theoretical framework to your issue), where do you discuss this and where do you discuss the actual interpretations that are given to some of your key concepts (that is, individual authors' delimitations in the field of meanings, or choices of definitions)? If you have opted to work from the "inside out", you will be appealing to different considerations raised in the literature throughout the thesis. So what then comes into the literature review chapter? Also, when you get to the section in which you analyse the data you have collected, you will usually do so in terms of criteria that have been developed in previous studies (written up in the literature). So do you discuss these in the literature section and then repeat these in your chapter dealing with analysis, or do you do another kind of literature review of data analysis in your analysis section?

These are tricky questions to which there is no set recipe answer – the guiding rule is that, whatever you decide on, it must be appropriate to the purpose of your thesis. Here are some suggestions about how you might think about deciding what goes where.

Many theses don't have a chapter called "Literature review", but draw on relevant literature throughout the thesis. However, these theses often do have a section at the beginning to sketch the academic background, dominant trends in the debates, current definitions, and so on, of the issue that the thesis addresses. This initial chapter draws on the literature pertinent to the topic (X) in order to sketch the academic context. The chapter might be entitled "Dominant trends in the academic debates on X" or "Theories of X" or "Current contestations in X" or "Main issues for consideration in X". Later sections in the thesis that address "foregrounded" issues, or the focus of the thesis, will then draw on specific literature to develop the details of the discussion.

See section 2.12, Appendix 3.1 and Appendix 15.

Other theses, again, do have a separate chapter on "Literature review". If yours does, then you might want to include the following in this: a summary of what the main approaches to your topic have been or are, what have emerged as key issues, how various authors have addressed these issues, what the main findings have been in the field, what some of the points of agreement and disagreement are, and how your thesis topic is located within these debates. You might want to consider grouping texts pertinent to "developed contexts" and "developing contexts" separately. Sub-headings assist you to structure the chapter and to group the texts together for a purpose. Remember, you're not already trying to answer all the questions, solve all the problems, or tackle all the issues in your first chapter – you are merely giving a solid background and base to the rest of your thesis, which will pursue more detailed discussion in subsequent chapters. You can work in a variety of ways.

See section 6.1.2.

Here are two suggestions of how you might structure this initial chapter. Start off with a broad context and then narrow it down in each subsequent sub-section. For example, a thesis topic on "The demographics of time use in urban South Africa" might have in its literature review chapter the following sections:

1. Overview of historical changes in time use
    1.1 In industrialised countries
    1.2 In post-industrialised countries
2. Interpreting the emerging pressures on time use
3. Research findings of paid time use and unpaid time use

Alternatively, list the most important key concepts that the thesis will address and work with. Then ask yourself the question: "What does the literature say about each one of these concepts?" Then, for each of these concepts, ask: "What are the dominant definitions or interpretations of this concept?", "What are the agreements and disagreements?", "What are some of the main problems?". For example, the literature review chapter of the thesis on "The demographics of time use in urban South Africa" could discuss how various authors have addressed the following key issues:

55

1. Various research findings on predictors of time use
2. Differences of time use in affluent and economically depressed contexts
3. Gender differences in time use
4. Age gaps in time use
5. Status and racial differences in time use

Note that the above structure for a literature review chapter could also form the overall chapter outline of a theoretical thesis in which each of the above sections is discussed in great detail in separate chapters, with each chapter drawing extensively on literature throughout the thesis.

There is a kind of reciprocal development between reading, writing and structure. The more you read, the more you discover issues you may not have anticipated or thought of. This can lead to puzzlement about where to fit in the new reading or consideration, and perhaps even about how to restructure the sections of the chapter. You have already read quite a lot in the development of your proposal, so the chances are that you have a pretty good idea of at least the broad background of and main issues in the topic. The purpose of this initial chapter is often to give an outline of the main debates, to give a clear and honest exposition of the main trends, to identify the key issues, to signal the concepts in terms of which you plan to analyse your empirical data, and to give an academic context to your study. Subsequent chapters will focus more on the details within your foregrounded area.

The question that drives your literature review chapter is always "What does the literature say about X?" At this stage of your thesis, your own response to X is not yet developed. However, in the thesis you need to demonstrate to your academic reader that your own response is *informed* by the literature on the topic. Before you can do so, however, you need to show that you understand what the literature is in fact saying. (You can't criticise or engage with something if you don't know what it is that you're criticising or engaging with!)

Here is a suggestion: next to each section of the outline of your chapter, identify about three key readings. You might find that one reading fits under a number of your sub-headings. Then you need to decide where you are going to discuss this particular reading. One approach is to summarise only those ideas that apply to the particular section of the chapter; discuss the same author in subsequent sections, again just summarising the ideas pertinent to that particular section.

If the perspective from which you are going to investigate is conceptually linked to the kind of method you will employ, then you might consider including the discussion about the research practices in the field of study in your literature review chapter. My own preference, however, is to encourage students to think about rather including this discussion in the methodology section (if the thesis chapter outline is so structured). In that way you are able to focus on specific aspects without becoming entangled in too complex a net. Since your whole thesis must be informed by the literature, you should link every chapter with considerations raised in the literature. It's not a matter of writing a literature review chapter and then ignoring the literature for the rest of the thesis!

4.1.4 Writing the first draft

Remember the two golden rules about writing:

1. Write early and write often.

2. Don't try first to get it right; get it written.

Putting your thoughts on paper (or a computer screen) helps to clarify your thinking and structure your discussion. Every author writes a draft, edits and re-edits. The slick, flowing texts you read in books and journals are products of long, hard work and polishing. Everyone struggles to articulate his or her thoughts clearly. However, like an athlete training for a marathon by doing short training runs every day, a writer becomes writing "fit" by writing something every day – the more you do it, the easier it becomes.

So you have a skeleton of your literature review or literature background section. And you have identified various key texts to consult for each section, through your own initiative by looking up references in the bibliographies of relevant journal articles and books, through reliable academic search engines on the Internet, through discussion with your supervisor and with your fellow students. Now you need to read the text closely for main claims, arguments, findings, interpretations, and so on. Your reader has probably not read the article so you need to give an honest and fair exposition of the author's ideas (whether you agree with them or not; you need at least to say what they are). For each reading, make an analytical summary. Once you have all the summaries of the texts pertinent to a particular section, read them all again and try to identify who is responding to whom on what issue, what are the points of similarity and difference, what are the main contentious issues, what are some of the gaps, and perhaps some of the implications of the main claims. Then group the summaries together in a logical sequence, with bridging paragraphs that highlight the links between the various texts you are discussing. Do this for each section of your chapter and you'll have the first draft in hand.

In summary:
- Develop right from the start a simple, yet rigorous system whereby you keep a record of all your references.
- Decide on an appropriate referencing convention.
- Ask your supervisor and fellow students for suggestions on your reading list of easily available books.
- Meet with the specialist librarian in your discipline.
- Know what constitutes plagiarism and how to avoid it.
- Refer to the outline of your chapters as developed in your proposal, revisit the sub-headings or themes in the light of new readings that you come across.

- Identify key texts for each sub-heading; one text may appear under a number of sub-headings.
- Write a honest and fair exposition of each reading, its main claims, findings, argument, approach, etc.
- Link the various coherent accounts you have written by drawing out differences, similarities, dominant trends, possible implications, etc., making sure that the purpose of grouping them together is clearly evident.
- Throughout the thesis be constantly on the lookout for relevant literature to incorporate in your thesis and inform your thinking.

4.2 EMPIRICAL DATA COLLECTION

In theses that pursue empirical research, the rigour of data collection is a key feature of the research process and may be one of the most problematic areas of the project. You must put good quality data on the bones of the initial research design developed in your proposal. Data collection is unpredictable and it is therefore important that both you and your supervisor remain flexible in your general approach to the project. Data collection, at the best of times, can be a protracted process. It is often labour intensive and time consuming. Students in general, unless they are part of a research team, have limited time and restricted resources. You must guard against a plan for data collection that is too ambitious – if in doubt, ask your supervisor for advice about what constitutes an adequate and realistic plan. All too often students plan to collect data that would take a medium-sized research team, with funded support, a fair time to complete! As Delamont et al. note (1997: 69), "Ambition is commendable, but over-optimistic estimates of time and effort are not to be encouraged". Think realistically and pragmatically about the scope of the data collection. Putting realistic limits on it can prevent you from "drowning" in data later on. Often over-ambitious data collection is a result of the student (and supervisor!) not feeling sufficiently *confident* to know when enough is enough. You both perhaps want to err on the side of over-collection, but this can lead to significant problems later on when the data must be analysed. Frame the plan for the data collection, bearing in mind the significance of the research questions to be addressed, the resources available and your own analytical capacities.

Students often ask, "How many informants should I interview?" Of course, there is no one answer to this, but a useful response to bear in mind is, "It depends on what you want to do". A modest volume of high-quality data analysed in considerable depth and with methodological precision will often be far better than a lot of data superficially analysed. Delamont (1997: 70) tells the story of Harry Wolcott, an American anthropologist whose approach was to study only *one* of anything at any given time: one village, or one school, or one community leader. People would often ask, "But Harry, what can you learn from just one?" to which

Harry replied, "As much as I can". However, it is difficult to know "just the right amount or right number" if the research is mainly exploratory and open-ended. The more data, often the messier it will be.

However, if you are doing surveys and statistical analysis, you need enough respondents to give your findings statistical significance. Given your research purpose, an expert statistician should be able to advise you on an appropriate sample size.

During the early period of empirical work especially, it is important for you to maintain your enthusiasm and confidence. If things were easy and research questions soluble as soon as we approached them, then research would be a lot more straightforward than it is. But almost always, research is hard and messy. For students in the social sciences, there is often disappointment that the social worlds they study do not readily match up with analytical concepts. You may find that you look in vain for your cherished ideas (such as "hegemony", "empowerment", "community", etc.), only to discover that the complex reality of the social world is not neatly organised in such terms. You may collect data, but cannot "see" a pattern or problem emerging, and you become insecure and fretful. Or if you have embarked on an impact study, you will probably find that there is no neat causal linear relationship between variables; behaviour is shaped by and responds to a multitude of different aspects, many unseen and unknown, in a dynamic, ever-changing relationship. Students doing laboratory work also often encounter experiments that do not yield their expected results, equipment that goes wrong, and failures in replicating findings. This is usually in contrast to your undergraduate laboratory experience, when the experiments followed well-trodden paths and yielded consistent results. That's why the experiments were chosen. The realisation that the outcomes of laboratory work are by no means certain may lead to a growing anxiety that you might not be able to meet your Master's or Doctoral degree requirements. There are a number of strategies through which you can learn to rationalise initial failure: realise that failure is not personal; it happens to everyone. Moreover, you can come to terms with failure by seeing it as a fundamental component in scientific research training, which is ultimately resolvable. As one experienced researcher expressed it: "Everything goes wrong but you have to remember that it's not all the time."

If necessary, in the face of persistent failure, you and your supervisor might want to undertake a realistic appraisal of the situation, review what has been learnt from the project to date, and apply those insights into a reformulation of the research question and research design.

The accumulation of lots of data can provide you with a kind of "security blanket" – the idea that the more data the better, that nothing will be left out, and that while the collection is going on, the dreaded analysis part can be deferred. You can create for yourself the sense of being very busy with the research, but this can often hamper the ultimate completion of the thesis. Too much data to analyse might become so difficult a task that you lose heart. Often in a situation like this you'll have to rely on your supervisor to help you achieve some sense of judgment of when enough is enough. In the case of "over-collection", you will need guidance on what is important and what can be discarded. Data can become precious

Van Schaik
Publishers

to you (all that hard work) and you might be reluctant to discard some of it. Don't think of discarded data as wasted data. You might revisit the data and aim perhaps for a journal publication once the thesis is safely in the hands of the examiners. In order to help you step back from the process every now and then, arrange to give a seminar presentation in the department, or write a conference paper, in order to refocus again on what is important for the thesis. Stepping back also helps you to see emerging patterns or trends which can escape your notice while you are immersed in the details.

> **In summary:**
> - Maintain a flexible approach to the data-collection process.
> - Guard against over-ambitious, time-consuming data collection projects.
> - Try to form some judgment of when enough is enough.
> - Maintain a productive balance between optimism and realism.
> - Remind yourself that perseverance carries rewards.
> - Recognise when ongoing data collection becomes a delaying tactic for completing the thesis.
> - Step back every now and then from the day-to-day data collection slog.
> - Distinguish between pertinent and other data – put "surplus" data on the back burner for the time being.

4.3 WRITING

Academics often tend to overestimate their students' familiarity with academic language and writing. So, if you're not sure what is expected of you when you're asked to "critically discuss", *ask* your supervisor. If some of the comments are not clear, ask. Remember, though, that even the most successful authors don't work out everything in their heads first and then write it all down lucidly and fluently. You discover what you think only through articulating it in words (preferably in written form). Also, reading can become the thief of writing. Reading can easily become a displacement activity, giving you the notion of being busy with the thesis, but not really producing anything concrete. The belief that one should *read first* before writing can put the thesis off indefinitely. Constantly bear in mind that reading and writing are reciprocal activities, the one informing the other. What you read informs what you write; what you write informs what you should read. Of course, your own and your supervisor's judgment needs to be exercised here. Some students write copious amounts with no reading of the academic debates; others read so much that it paralyses them for writing. In the first case, draw up a definite reading list and then work through it by writing summaries or commentaries on each reading. In the second case, agree with your supervisor on a specific and modest writing task in order to break the fear of writing.

Writing involves management of time and place. Every person has different rhythms of times when he or she is most alert and motivated. It is important that you establish a *routine* of writing time and place. Hope that the writing will be slotted into "gaps" in your general schedule of family and work commitments is often ill-founded. More often than not, the writing is postponed and, the more it is postponed, the more difficult it is to get started again. Like a car that has come to a standstill, more petrol is needed to get it moving again, as opposed to a car that is already chugging along. Commit yourself to a routine that works best for you. Some prefer working early in the morning, others prefer late at night; some prefer working from home, others prefer another place. A colleague of mine wrote her entire thesis in various coffee shops. She said the background bustle and the fact that no one could phone her there were conditions conducive to her writing. Some like working in silence, others prefer background music or noise. Whatever the conditions, discover what works best for you and set up those conditions. (Many universities have small carrels available on campus for Master's and Doctoral students – find out from your supervisor or from the librarian.)

If you are struggling with language, there are various measures you can take: enrol for a language course (most universities offer these); find out if there are academic development services available on campus and use them; join (or start) a reading and writing group with fellow students; speak, read and write as often as you can; appoint a professional proofreader and editor to go over your written text with you; learn to use the spell and grammar checks on the computer.

Thinking about the overall thesis can be daunting, so break up the writing of the thesis into manageable bits. I have often got my students to frame each chapter topic as an assignment question. If the thesis has, say, ten chapters, I encourage students to see each chapter as a 10–15 page assignment with relevant readings to be consulted. (Of course, the fewer the chapters, the longer they will be.) Writing assignments is something you have done in your Honours year at least. Once all the "assignments" have been written, the body of the thesis is there. Then linking paragraphs need to be inserted so that all the "assignments" become one coherent extended argument. And there is the thesis (well, almost).

Many writers at times experience "writer's block". When this happens you can try one of the following "unblocking" moves:

- Write *around* missing information – instead of trying to find that missing reference, just put in the text "????" to remind yourself to chase up the reference at a later stage.

- Be realistic in the writing tasks you set yourself – one page is much better than no pages at all.

- Set up conditions conducive for writing – arrange a suitable work space, time with no interruptions, avoid having a heavy meal or alcohol before sitting down to write (they will make you feel sleepy!), and deal constructively with what's blocking you, for example, by talking with friends or colleagues about the section that has you stumped.

- Another strategy that might work is to "talk" through the chapter or section with your supervisor during your meetings. Tape the session and later, when listening to the recording, jot down the points you have made and then develop these as a framework for a written piece.

- Write on *anything* for ten minutes – just to get going.
- If the worst comes to the worst and, despite all your best efforts, you are not producing, you need to act quickly. You can readily recognise writer's block when you find that you are postponing appointments, pleading lack of time, or taking refuge in reading as a displacement strategy. Perhaps it's time to get tough on yourself. Set some definite deadlines with your supervisor, ensure that you produce small bits of writing for each one, register for a conference, or arrange to present something at a departmental seminar. If you still fail to produce, it is time to look closely and honestly at what is holding you up. It could be fear of failure, a misguided perfectionism, unhelpful working habits, or apprehension of your supervisor's criticism.

A reminder: produce your text on a computer. Buy or loan a PC if you don't have one. Word processing allows you to play with the text and move it around. This loosens the grip of the text from becoming too precious to let go (and saves a great deal of time).

In summary:
- Write early and write often.
- Dispel the belief that writing is a separate kind of activity that can be left till the end (as in "writing up" the research!).
- Create conducive conditions for writing.
- Set clear, modest writing tasks in the early stages.
- If necessary, use the academic support services offered by the university.
- Break up the writing of the thesis into manageable bits.
- Deal with writer's block by changing tactics or by an honest appraisal of the reasons holding you up.
- Use a computer for writing your text.

See Appendix 10 for some practical suggestions for thesis writing.

4.4 KEEPING UP THE MOTIVATION

Almost every student writing a thesis goes through dips of depression, insecurity about the quality of the writing, anxiety about the thesis requirements and non-thesis related problems that nevertheless impact on the thesis production. In the beginning of the process especially, you may still be trying to find your "academic feet". Indeed, at many stages of the thesis writing, you'll need to be reassured and motivated. Bear in mind that the main task of a supervisor is, of course, to give clear, crisp and top-level professional academic advice, so don't feel personally devastated by sharp feedback. It is easy for one's emerging academic self-confidence to be destroyed. If there is a section with which you are struggling, with an increasing sense of failure, suggest to your supervisor that you will revisit the shaky parts later on when you have read more and have had some distance from what is becoming a demotivating experience. Once you have invested more in

© Van Schaik Publishers

the success of the thesis, you can revisit the problematic parts. Of course, your supervisor shouldn't blindly encourage you if your work is clearly inadequate or if you are heading in quite the wrong direction. But don't take this feedback as a personal affront. Examiners ultimately examine the product only – they know nothing about how nice a person you are, or how hard you have worked, or how important it is for you to qualify for promotion purposes. Your supervisor would be betraying you if she were to express approval for work that is obviously below acceptable standard.

4.4.1 Student responses to the thesis

Smith (2000: 240–255) notes various stages in student responses to thesis production. Knowing what these responses are and being able to recognise them should help you guard against becoming trapped in them:

Enthusiasm

In the early stages of thesis writing, this is often revealed in your trying to do too much and developing over-ambitious proposals. However, as time progresses, as the need for rigorous writing becomes pertinent, as the stress of time constraints and the monotony of focusing on a particular issue are more keenly felt, enthusiasm can wane dangerously. Maintain enthusiasm by:

- joining or establishing a reading and writing group with fellow students.

Isolation

For Master's students especially, who have completed the coursework part and are suddenly on their own with no more structured class meetings, little interaction with fellow students and no given course outline, the sudden isolation and lack of structured routine can be difficult. Part-time students too, including Doctoral ones, may find that the limited opportunity for intellectual stimulation and exchange of ideas can lead to a loss of interest in the work, coupled with a markedly diminishing momentum. Use the following strategies to combat isolation:

- Ensure regular meetings with your supervisor.
- Join or set up a Master's and Doctoral reading and writing group. Meet regularly.
- Attend departmental lunches or socials if you may.
- Set up an email network between all the postgraduate students in the department.
- Ask to present a paper at the departmental or faculty seminar forum.
- Register for a conference.
- Connect with someone writing a thesis on the same topic. Consult the library database for a list of current registered Master's and Doctoral titles.
- Find out if there are employment opportunities in the department, perhaps as a tutor.

Boredom or loss of motivation

Recognise what the possible origin of this might be: a distaste for a specific aspect of the work (for example the analysis part, writing, finding a site); a temporary loss of enthusiasm for the whole task; or a serious, perhaps clinical, depression. To work on the same thesis topic for an extended period can become monotonous and repetitive. If the work is proceeding as it should, you may lapse into a sense of the thesis being too predictable. Alternatively, you may suffer from the "getting nowhere" syndrome. Either way, you could tackle it this way:

- Draw up a weekly plan for systematic reading, working and writing of tasks.
- Talk to others about your thesis – others who don't know about your topic. Their interest in the topic could re-inject you with a measure of enthusiasm as well.
- In the case of a distaste for a particular section, put that section aside for the time being and work on something else until you are remotivated. Start other parts of the thesis alongside the less preferred one.
- Realise that one reason that you may stall at the "writing up" stage is that you have left yourself no other tasks but writing. If you had written while doing some of the other parts, then it would not be so "all or nothing" at the end.
- In the case of a temporary loss of enthusiasm, keep in regular contact with your supervisor and fellow students. This might help to recover your initial motivation.
- Consider going to a conference (not necessarily to present) or attend a faculty seminar.
- Contact other students working on the same topic.
- In the case of serious depression, get access to professional care. Consulting your Student Health or Student Counselling Unit could be a fruitful first step.
- Consider taking a few days' break and go away somewhere with your family and no thesis work; relax, replenish your resources and rekindle your motivation.

Frustration and anxiety

As the work progresses, there are often new avenues opening up for further exploration. It is tempting to pursue some of these, but that would mean delaying the completion of the thesis within the allotted time. It becomes important to maintain clarity about the end goal, the main research question and the procedures designed to address it. Keeping a focus and resisting being sidetracked can become more and more difficult as the original problem becomes more and more familiar. Some theses are structured in such a way that the writing up part comes at the end. Most students radically underestimate the time and effort this stage demands. Having surveyed the field, collected the data and analysed the findings, you may think that the writing up is going to be a merely technical undertaking, fairly easily accomplished. Not so. Writing up usually demands the most concentrated effort of the whole thesis production. And it is therefore at this stage that you are at risk. You may struggle with this stage for a number of reasons. You may feel that all the "real" work has been done and there may be little motivation left

64

to go and revisit the work. Moreover, at this stage you may find that you have ambivalent feelings about the whole project and may be tempted to run away from it all, now that the data is actually there for others to see. Another reason you may be struggling with this phase is that, unless everything has gone remarkably smoothly (it overwhelmingly doesn't!), you may have to make some major adjustments in your arguments or reinterpret some of the findings and revisit the presentation of the data. This calls for academic competence and professional discipline. And lastly, you (and many other students like you) struggle at this stage because most of you have never written anything as long as a thesis before, and persevering through the task takes considerable effort. You need to:

- keep the overall outline and common thread of the thesis clearly in mind
- step back from the details every now and then
- remind yourself that countless others before you have managed to complete their thesis, so it is "doable".

Pressure to finish

One of the most pervasive emotional responses to thesis production is that of anxiety. Throughout the various stages of thesis production, anxieties about finishing it are present. Eventually, the thesis is seen as something that you must finish. Often towards the final stages, many students long to be free from this constant nagging anxiety. Here you need to:

- go for closure
- keep a close eye on potential lapses in rigour as you rush to finish.

Separation anxiety

While some students can't wait to get rid of the thesis, others close to the completion of the thesis suffer from separation anxiety. Often this manifests itself in your making numerous unwarranted and insignificant editorial changes, questioning the validity of your work, and wanting to incorporate additional readings. Here you need to:

- rely on the academic judgment of your supervisor and be firm in going for closure if the standard of work is appropriate
- draw the boundaries clearly. (By definition, all research is ongoing. There are always more books to read and more ideas to consider.)

Depression

This is often a "post-thesis" syndrome that sets in after the thesis has been submitted for examination and you await the result. After having had the thesis dominate your life for a number of years, usually accompanied by a frantic work period towards the end, suddenly there is nothing. The loss of the thesis routine and the anxiety while waiting for the results can lead to post-thesis blues. Remind yourself that:

- there is nothing you can do at this stage about the thesis

- it might be an idea to look at parts of the thesis that could potentially be reworked for publication in a journal.

Closure

Once the changes recommended by the examiners have been received, you can actively start to get closure on the thesis. Ensure that the correct procedures are followed and that you fulfil all requirements (pay up fees, return all library books, hand in final copies, etc.) in order to be able to graduate. The celebration of the completion at the graduation ceremony is an important public acknowledgment of your achievement. Enjoy the accolade! You need to:

- follow the university procedures carefully. Not fulfilling the necessary requirements can lead to a delay in your graduation and bitter disappointment.

• •

4.5 ADDRESSING RISKS TO THE COMPLETION OF THE THESIS

Inadequate supervision is often cited as a reason for students dropping out of their studies. Penalties for dropping out are much greater for you than for the supervisor and it is often up to the determined student to insist on adequate supervision. Clarifying the ground rules at the start of the supervision process can eliminate such problems later on. Most universities have procedures in place for student complaints if you find that you can't work productively with your supervisor. If necessary, find out what these are and provide documentary evidence on which you base your complaint or request for a change of supervisor.

Bear in mind that when you first applied for entry to Master's or Doctoral studies, you went through a selection process and the department obviously judged you to be competent to complete the degree. Live up to it. However, obstacles to thesis completion are not always of an intellectual kind. Sometimes there are obstacles in the social environment. These could include problems such as lack of money, poor working habits, not being able to juggle thesis work with family commitments and job demands, or a new job that has been taken on before the thesis is completed. Possible ways in which you can address these are, at the simplest:

- Have a honest and open discussion with your supervisor about the problems you are trying to deal with.

See Appendix 19 for sources of funding.

- In cases of lack of money, find out if you can secure a part-time job in the department, or can access some funding from a research project or apply for a scholarship.

- In cases of inappropriate work habits, ask others what works for them (don't think that others don't have similar problems). Think about conditions that work best for you.

- Instead of trying to fit your thesis work into the few gaps you have between work and family, organise your work and family around your thesis. (When I was writing my thesis I periodically pencilled in a 2–3 hour meeting with myself in my diary!)

4.6 PROGRESS REPORTS

Some universities require progress reports. This may seem like over-bureaucrati-sation, but it has the important function of forcing you and your supervisor to step back regularly and to assess progress made. It can serve as a handy signal that you are approaching a risk situation or, alternatively, it can serve as a great motivator. You may think that you have made little progress since you are so immersed in the research, but stepping back and having to note what you have done over a certain period may lead to you being quite surprised at just how much you have been able to accomplish! Even if your department does not require a progress report, you may want to suggest to your supervisor that the two of you review your progress every six months or so.

Developing academic discernment

Over the period of your postgraduate studies, you are expected to develop the academic equivalent of "good taste". During this time, you need to learn to judge your own work by standards appropriate to postgraduate work and shared by the community of scholars in the specific subject field. This is not a "technical" skill like proper referencing or experiment design. Instead, it is something much more implicit and indeterminate. Learning to judge research in the subject field is something that develops over many years through active involvement – be it through reading or networking – with participants of that academic community. Students just starting out in their studies are often unsure about their own ability to judge the work, especially whether it is "good enough" to be submitted for examination. For that one needs to have academic discernment. One of the central elements of academic discernment is an appropriate sense of critical engagement. But what does that mean?

5.1 CRITICAL READING, THINKING AND WRITING

How often have you come across the following instruction: "Critically discuss ..."? It is a phrase used often, but not always with a clear *understanding* of what it means to discuss something critically. I once asked my colleagues who had set exam questions starting with "Critically discuss ..." what exactly they were expecting their students to do. After some initial vague responses, such as "Well, you know, *critically* discuss", I pushed them to spell out exactly what they thought this entailed. The responses varied widely. Each colleague had a specific idea of what he or she thought it meant: "Well, students have to say why they agree or disagree with the author"; "Students must analyse the structure of the argument, assess its validity and determine the truth of the claims"; "I expect students to highlight the underlying assumptions the author makes"; "It means they have to contextualise the author's claim"; "I want my students to develop the author's main idea further by examining the possible implications of the claim"; "What I am looking for is whether students have analysed the meanings of the concepts used by the author"; "Obviously it means that students must assess the contribution the author makes to the existing understanding of the topic", and more! Now, each one of these is a pretty sophisticated academic skill. Are you expected to do all of these? What does your supervisor expect you to do? What is the accepted academic expectation in your subject area of what critical discussion entails? The following are guidelines to help you find some structure when developing a critical discussion, and at the same time growing academic discernment.

More often than not, an initial response to the instruction "Critically discuss ..." is to think that we must find fault, or highlight the weakness in the argument, or reject certain claims. Although this might at times be part of a critical discussion, it is by no means the *only* or even *most appropriate* way to engage with the claims expressed. So, before we look at what critical discussion (or engagement) is, let's get clarity about what it is not:

- Critical engagement is not the same as disagreement.
- Critical engagement does not aim to embarrass, humiliate or seek to dominate.
- Critical engagement does not mean nit-picking.

So, what *does* it mean to read, or think, or write critically? I've noted that critical engagement (which incorporates critical reading, thinking and writing) is *not* merely rejecting or finding fault with someone's argument or position. Rather, it is *a rational reflection on one's own and others' ideas in order to get a clearer understanding of an issue.* The following are some pointers to assist you both in your reading of others' texts and in constructing your own writing for your thesis.

- Critical engagement means giving a **clear exposition** of the argument.
- It entails determining and assessing the **support** for a certain claim that you or others have made in order to get a clearer understanding of an issue.
- It means determining the **truth** of the premises and the **validity** of the argument.
- It entails clarifying and analysing the **language** used and the meanings of concepts.
- It involves showing how the article or book fits into the academic debates and **current literature** – to whom or what is the text responding?
- It entails discussing the theoretical and social **context** in which the ideas are developed.
- It involves a discussion of the possible **implications** that the ideas or claims could have.
- It demands **informed thinking** and **creativity**.

One of the main things to remember when engaged in critical reasoning is that it is a two-step process: first, you must have a clear understanding of *what* the author is saying, before you can, second, evaluate the ideas expressed. You can't criticise, reject or accept something if you don't know what it is you're criticising, rejecting or accepting. Too often students skip the first step and launch straight into their own passionately held opinions. This is not critical engagement, unless you have demonstrated that you understand the debate before you contribute your own ideas to the discussion. Figure 5.1 represents critical engagement graphically.

Figure 5.1 Aspects of critical engagement

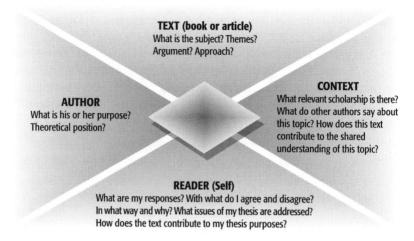

Source: Adapted from Taylor (1996: 6).

Here are some steps you can follow in order to systematically develop the task of first clarifying the author's ideas and then shaping your own critical engagement with these ideas.

5.1.1 Exposition of the text

1. Pre-read

One way of getting a quick idea of the main thrust of the text that you are going to engage with is to do some pre-reading before starting on the actual text:

- Read the flyleaf of the book (what is the main idea the author is developing? Who is the author? Where is she located? When was the text written?).
- Read the preface or introduction of the book (often in an edited collection there is a short analysis of the various chapters).
- Read the abstract (journal articles have a summary of the article at the beginning).
- Check the reference list (are current, important works listed?).

2. Read and reread the text

Don't expect that you will grasp the ideas in the text on one reading only; this usually involves a number of rereadings.

- Do you understand the literal meanings of the words used? Keep a dictionary at hand and look up key words whose meanings you're not sure of.
- What metaphorical meanings (that is, meanings through association) does the author develop?
- Are there cultural terms or idiomatic expressions you're not familiar with? A good dictionary will spell these out.

3. Make systematic notes

One of the ways to find out what you are thinking about what the author is saying is to write down your thoughts. It is extremely important to work towards seeing a clear and accurate picture of the text. Suspend your judgment for a while, focusing instead on outlining the text. A student once described this as "listening to the author's voice rather than to my own". Remember, the first step in critical engagement is to clarify the author's findings or argument for your reader.

In order to help you analyse the content and argument of the text, number and read each section or paragraph carefully and then do two things: jot down its main point or idea, and then its function in the text – or, putting it more simply, write down in one sentence what each section *says* and what it *does*. Here is an example:

The case for brands

In many parts of the developing world, one still finds little presence of brands. That is, one finds the sort of environment the anti-brand campaigners say they prefer. No billboards advertising Castle, no McDonald's, no annoying radio jingles. However, whenever wealth increases, the presence of multinationals as investors, and the products of multinationals for consumers, are eagerly sought. With them of course come the advertising of brands. Should we suppose that these people who happily ditch their local beer for a branded one and discard their homemade sandals for Nikes are all dupes of international capitalists? Or should we conclude that there's something brands provide which people, in general, prefer having to not having?

Naomi Klein is the best-known proponent of the idea that people's tastes are enslaved by brands. Her book *No Logo* is the leading statement of this view, making up in passion what it lacks in logic. According to Ms Klein's pawn theory, brands are the symbols and flagships of a "fascist state where we all salute the logo and have little opportunity for criticism because our newspapers, television stations, Internet servers, streets and retail spaces are all controlled by multinational corporate interests". Brands are too omnipresent and powerful to allow genuine choice, she claims. In addition to branded goods often being manufactured in third-world 'sweatshops', they also promote a dull and homogenous world from which genuine local colour and tradition are driven out.

This entirely ignores both the history and economic function of brands. They did not arise, in the first place, to tyrannise consumers but to protect them. Before the industrial revolution, people didn't need brands, because most goods were produced locally by people whom the buyers knew. Once people congregated in large cities, the guarantees provided by such direct familiarity were no longer available. Brands took over this function. The owner of a brand has a powerful incentive to make sure that the quality standards associated with it are not allowed to slip or vary from place to place and must remain consistent over time. Brands have therefore grown in importance and visibility along with, and as a function of, the increase in global interconnectedness. This has of course increased their value – and, with it, their owners' stakes in making sure that they really are indicators of reliable quality.

Source: Adapted from *The Economist*, 8 September 2001, p. 9.

Make two columns:

Par.	What does the author say/ask?	What does the author do?
1	Do brands make pawns of people or do they offer them something better than they had before?	Introduces the two sides of the problem question
2	Klein's pawn theory holds that brands disempower consumers and curtail choice.	Summarises the main claims of the pawn theory
3	Brands protect consumers by providing a guarantee of reliability and quality.	Introduces a historical perspective to substantiate an alternative view.

4. Examine, categorise and summarise your notes

Once you have done this, read through your notes and then write down your responses to some of the following questions (not all may be relevant to the specific text). This will help you develop a clear and honest reflection of the author's text.

- What is the **topic** that the author addresses? What is she specifically focusing on? What are the boundaries of the topic?

- What is said? What is the **main idea** that is developed? The main claims? What is the author's main position with regard to the topic? Where does she stand? (See the "tree" structure, later in this section, to help you determine the author's main idea.)

- What is the **structure** of the text? What comes first, second, third? Why this sequence?

- What **interpretations** are offered of the main concepts? What does the author mean by "x"?

- How are these main ideas or the position supported? **Support** can be offered in various forms: references to other authors; examples, case studies; metaphors – using an image to illustrate a point; reasons and development of argument; conceptual analysis; cause and effect; statistics; literature review; historical contextualisation.

- In which **theoretical framework** is the author located? Through which conceptual lenses is the author looking at the issue? Remember, there is no such thing as a "neutral" view.

- What **methodology** is the author using in order to make sense of the issue or to develop a particular position?

- In what **context** is the author writing? Place? Date? In what discipline? In response to what?

5.1.2 Evaluation of the text

5. Assess the main idea, the structure of the argument, the author's purpose, the context, the contribution to scholarship, and possible implications of the findings

By now you should have a clear grasp of what the author is saying. This is a neces-sary first step in reading critically, but it's not enough. Now you need to evaluate what the author is saying. Remember, critical engagement does not mean nit-picking, nor does it necessarily mean that you have to find fault; rather it means that you assess the author's contribution to the ongoing academic discussion about the issue. In order to get a grasp of this, develop a response to some (or all) of the following questions:

- Are the **limits of the topic** appropriate? What is not said? Why not? Is this a seri-ous omission? Are the limits of the topic too narrow, too broad? If the limits are too broad, is the author in danger of generalising too much? If the limits are too narrow, is the author saying anything of interest?

- Are the interpretations of the main concepts offered clear? Does the author rely on buzzwords or popular rhetoric, or are the **meanings of the key concepts** clearly stated? Words or concepts are not "discovered"; they are constructed for particular purposes. What are these purposes? Language embodies the perspective from which we view a particular issue. It reveals how we order our experiences, what assumptions we make, and reflects what we think. (Since ideas are expressed in language, the analysis of language is extremely impor-tant in our critical engagement with these ideas.)

- Are **supports for the main claims** appropriate to the context? (Does the author, for example, use findings, illustrations, or metaphors from other fields or disci-plines and transport them into another context? If so, are these illuminating or distorting? In what way?) Are the claims true? (On what facts and figures do the claims rest? Is there counter evidence that might contradict these supports?) Are the claims valid? (Does the author systematically develop the position or are there jumps in logic?)

- What could be possible **counter-examples**? What other perspectives or conclu-sions are possible?

- Does the author make certain **assumptions**? (For example, does the author assume a high literacy rate among the population, a certain level of economic welfare, certain divisions in society?) Are these assumptions justified? Of course, no author can spell out all the assumptions on which the ideas she develops are based, but you as a critical engager need to be able to judge whether there are assumptions that ought to have been spelt out but aren't.

- Is the **methodology used appropriate**? Could the author have followed another fitting methodology? What does the author's methodology (or the other one) bring to light that the other approach doesn't? Is the sample size appropri-ate? Are the criteria for choosing the sample appropriate for the main idea the author develops? Is there evidence of verifying the findings?

- How would the issue have been interpreted in another **theoretical framework**? (Think of the duck/rabbit example which illustrates that the same drawing can be interpreted in two quite different ways.) Can the claims or findings be inter-preted from a different perspective: from a gender perspective, an analysis of power relations, a post-modern stance, a materialist framework, a liberal stance?

- Is there anything of **relevance missing** from the book? Certain kinds of evidence, or methods of analysis/development? A particular theoretical approach? The experiences of certain groups? Remember, no text can address everything, so don't nit-pick on inevitable gaps that are not significant for the overall findings discussed by the author.

- Can you analyse the argument in terms of necessary and sufficient conditions?

- The ideas are expressed in a particular context (date, setting, discipline, in response to a particular issue). How could these be **extended** into perhaps another context? How could the ideas expressed by, for example, an American author be used fruitfully in a South African context? What are the particular dynamics of South Africa that would have an impact on these ideas? Also, how could ideas expressed by, for example, a political scientist be used in an educational setting? In other words, what would be the implications of these ideas in, perhaps, a school setting?

- What *experiences of the author* might have influenced her writing? What else has the author written? How does this work link with the author's previous writing?

- What are the **connections** between this text and other texts written on this topic? What contribution does this text make to the shared understanding of the subject? What is the scholarly or social significance of this text? Texts are part of an ongoing academic conversation. It is important that you have some idea of where in the academic conversation this text can be placed.

- The fact that the article or book has been published signals that it contributes something of significance. Why do others (and you) consider this to be an important text?

- You have selected this particular text (or topic) because it has "spoken" to you. Why? In what way? What questions would you ask this author if you could?

- What, for you, were the three or more best things about the text? The three or more worst things? Why?

6. Determining the author's main idea: a tree structure

It is often quite tricky to articulate in a short, clear sentence the main idea or claim or finding that the author is advancing. Since the whole purpose of a text is to compel the reader to accept a particular position, the main idea (or thesis statement) is really the one around which everything centres. Without such a centre, a piece of writing would be a muddle and pointless. So, if you're seriously interested in engaging critically with a text, you must first determine its main point or thesis. This is crucial, because everything written by the author can be assessed only in terms of the contribution made to the main point. Not all authors state the main thesis clearly. But *every academic text deals with some topic and its author always has an attitude toward that topic*. To determine the topic, go back to your notes on the paragraphs, and try to determine a common concern of the paragraphs. Also, consult the abstract, title or summary of the text – often the author will state the main claim there. Here is an example, adapted from Barry (1984: 302):

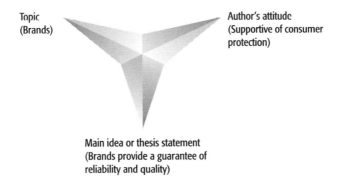

Topic
(Brands)

Author's attitude
(Supportive of consumer
protection)

Main idea or thesis statement
(Brands provide a guarantee of
reliability and quality)

5.2 A CRITICAL TAKE ON THE THEORETICAL FRAMEWORK AND METHOD

You need to be prepared to deal with the theoretical perspective and its adopted research methods within which you will be working, and to be intellectually committed to it and able to demonstrate command of it before the thesis is too far advanced to change it. This is particularly pertinent in the social sciences, where theoretical and epistemological issues are hotly contested. The choice of theoretical perspective (and research practices) determines the way the problem is interpreted, the approach to it, and what counts as a rigorous response. According to Delamont et al. (1997: 44), "graduate students become identified with such theoretical stances too early and too firmly, adopting them as articles of faith rather than subjecting them to critical scrutiny".

One cannot really speak of "choosing" a theoretical perspective. At the stage of postgraduate studies, it is likely that you have already become initiated into a particular perspective, have become committed to a certain academic style and have identified with a particular intellectual tradition, especially one that is prevalent in your specific subject area. Sometimes the working relationship between you and your supervisor is based on a commonality of perspectives which does not always make for critical engagement with the fundamental assumptions. Nevertheless, you should work critically within particular perspectives and resist adopting a given position while you remain ignorant of competing orientations.

The extent to which you need to become a critical inquirer of the chosen theoretical perspective, and to consider alternative approaches, depends a great deal on the type of thesis that you are writing. Master's minitheses are generally much more limited in scope and depth (nevertheless demonstrating academic rigour within a limited area) and need not necessarily offer a defensive argument for their choice of theoretical perspective. A minithesis can articulate clearly the assumptions on which it rests (with appropriate references) and proceed from these as starting points. However, in the development of the discussion, gaps or

Ⓒ Van Schaik
Publishers

weaknesses or contested parts of the theoretical perspective should be addressed or at least acknowledged. A Master's full thesis is expected to offer a more extensive discussion, including a justification for its adoption of a particular theoretical framework. A Doctoral thesis, generally regarded as a distinct contribution to knowledge of and insight into the subject, is expected to be more extensive and penetrating in discussion, including a critical awareness of the theoretical framework within which it is operating.

See Appendix 11 for blocks to critical engagement.

5.3 ACADEMIC PARTICIPATION

One of the enriching experiences of being a postgraduate student is feeling that you're part of an academic community. During your period of study, you should become familiar with how the scholarly community exercises its judgments. Actively participating in academic activities will help to motivate you and sustain the thesis production. It may also encourage you to think about pursuing an academic career. (Vibrant universities need up-and-coming scholars in the field to open up and continue areas of exploration.)

Being involved in the department's and the university's academic life is stimulating. There are various ways in which you can do so. There is substance to the old saying that if you really want to understand something, teach it to others. If there is tutoring work available in the department, you might be able to use the teaching space to engage with some of the issues relating to your own research work. Enquire whether there is a suitable research project you could join. Working with others in the same field can be academically stimulating. Or you may volunteer to present a paper, even if it is work-in-progress, at a "peer review" session or at a departmental seminar. You may also get involved in the broader academic community. Most professional societies or disciplinary associations have annual conferences and special membership rates for students. Ask for suggestions on which conferences to attend and which society to join (membership usually includes a subscription to an academic journal). Submit an abstract of a paper based on your thesis work for presentation at the society's conference. You might even think of co-presenting with your supervisor. Attending a conference not only gives you a deadline by which to produce a presentable text, but also opens up space for fruitful feedback from conference attendees, which can be incorporated into the thesis. Moreover, conferences are ideal opportunities for networking. Use the opportunity to get people's email addresses and to set up correspondence with them.

You might consider submitting a paper for publication in a journal. After having presented a paper at an academic conference, and after having reworked the paper in the light of feedback received from conference participants, identify a suitable journal to which you can submit your paper. Especially if you are a Doctoral student, you might think about presenting a series of conference papers, each one based on a chapter of your thesis. Doing so will give you useful feedback from members of the relevant academic community, enable you to network extensively, and let people know what you are working on. Once you have presented the paper, rework it for submission to an academic journal. Make sure that you adhere to the "instructions to authors" issued by the journal – these are usually published on the inner back cover of each edition. Hopefully you will get perceptive comments from the journal referees, which you may then incorporate into your chapters, and so strengthen the academic rigour of your thesis. If you have some of the articles published, this means that you both get feedback from the top researchers in the field and have a publication record by the time you have completed your thesis. Two birds with one stone!

In summary, take the initiative in extending your academic participation:
- Actively participate in the academic life of the department and university.
- Enquire about the possibility of tutoring work in the department.
- Find out if there is a suitable collective research project you could join.
- Volunteer to present a paper based on your thesis work at a "peer review" session or at a departmental seminar.
- Make a point of meeting visiting scholars in the department.
- Ask your supervisor to assist you in finding a suitable professional association to join (and to subscribe to the journal).
- If necessary, with your supervisor's help, submit an abstract of a paper for presentation (or co-presentation) at a conference.
- Network productively with other attendees at the conference or via email with scholars in the field.
- Rework the conference paper for submission to a journal.

5.4 THE END IN SIGHT

As mentioned earlier, the last stages of the thesis writing can often be the most demanding and difficult. You may have lost your initial enthusiasm for the project, may be somewhat bored with it, may want to start working on other research initiatives, and maybe you just want to have a bit of a social and family life back again. If you are feeling this way, take heart – many others have gone through the same stage and have nevertheless successfully completed their theses. At this stage it is common to find it difficult to summon the last spurt of academic energy to finish.

If you are tempted to abandon the entire enterprise, don't! At least, not before you have had a honest and open talk with your supervisor. Being so immersed in the thesis writing over the past few years, you may easily have lost the necessary academic distance to see its worth overall. Here you need to trust your supervisor's judgment or, if the supervisor is also new and not sure, appeal jointly to a more experienced academic in the department to make a judgment.

A Doctoral student said that what kept her going was the fact that she had been given a generous bursary, and not completing the thesis would mean that she would have to pay back all that money. She just didn't have it, so was "forced" to complete the work. Also, it might be psychologically useful to identify a role model – a student who has already completed his or her thesis, perhaps under difficult circumstances. Draw strength from the fact that if she could do it, then so can you! Ask your friends and family to push you through the last stages – instruct them to curtail your delaying moves and deferment tactics. Visualise yourself on that graduation stage, and the photograph of the occasion that you will be able to display. Hear your colleagues congratulate you or call you "Doctor". Take comfort in the unassailable truth that the sooner you get to work, the sooner the whole thesis will be done. No fairies come during the night and write the chapter for you – only you can. It takes a lot of self-discipline to work systematically on the last few details and sections. You don't want to jeopardise the whole enterprise by doing sloppy work on the last bits.

If, however, you are too anxious to allow yourself to let go, and find yourself constantly making editorial changes here and there, trying to incorporate yet another reading or reference, it is again time to have a honest and open discussion with your supervisor. Trust your supervisor's judgment when "enough is enough" and when the thesis is ready for submission. Remember, there will be lots of opportunity to rework sections of the thesis into journal articles, or maybe even the whole thesis into a book. It's not as though this will be the last word you'll ever write on the topic.

Part C

Preparing the thesis
for examination

Bear in mind the standard content length of the thesis. Accepted lengths differ from discipline to discipline. International standards for a Master's thesis are usually about 20 000 words, and for a Doctoral thesis 80 000 words. Although there are no hard and fast rules about the length, here are some indications of general practice:

- A research report: an average of 35 pages in 1.5-spacing
- A Master's minithesis: an average of 100 pages in 1.5-spacing
- A Master's full thesis: an average of 150 pages in 1.5-spacing
- A Doctoral thesis: anything over 250 pages sends a signal of the thesis being in danger of lacking focus and rigour.

Some information about the parts of the thesis

6.1 THE PARTS OF THE THESIS

Your thesis must have the following parts (unless they are marked optional below). The parts must adhere to the following order, unless alternatives are indicated:

Preliminary sections, pages numbered in roman numerals (i, ii, iii, iv, etc.):

> Title page
> Keywords
> Abstract
> Declaration
> Acknowledgements (optional)
> Contents
> List of figures, tables or acronyms (optional)
> Preface (optional)

Main text, pages numbered in Arabic numerals (1, 2, 3, 4, etc.),
> divided into numbered *chapters*, each with a heading

Endnotes (unless you use **Footnotes**)

Bibliography (or **References**)

Appendices (optional)

(These part names are used as headings in the thesis.)

● ●

6.1.1 Comments on the preliminary pages

Title page

The title page should carry the following particulars:

- The full (registered) title of the thesis
- The full names of the author

- The following statement:

  For a Master's minithesis:

  A minithesis submitted in partial fulfilment of the requirements for the degree of ... [Magister Educationis, Magister Artium, Magister Scientiae, Magister Philosophiae, etc. – as applicable] in the Department of ... [as applicable], ... University.

  For a full Master's thesis:

  A thesis submitted in fulfilment of the requirements for the degree of ... [Magister Educationis, Magister Artium, Magister Scientiae, Magister Philosophiae, etc. – as applicable] in the Department of ... [as applicable], ... University.

  For a Doctoral thesis:

  A thesis submitted in fulfilment of the requirements for the degree of ... [Doctor Educationis, Doctor Philosophiae, etc. – as applicable] in the Department ... [as applicable], ... University.

- Date submitted for examination (month and year)
- Name(s) of supervisor(s) (optional)

See Appendix 12 for an example.

(Examination copies should have a heading "Examination Copy" on the title page.)

• •

Keywords

In order for your thesis to be taken up in a library and research data base, you need to provide *ten* keywords (or, where unavoidable, compound words or phrases) that capture the main concepts of your thesis. The keywords should give someone who has not read your thesis a clear idea of the issues you tackle. Ensure that you sequence the keywords in a logical order, one that reflects the development of your argument. Keywords can also be included as part of the abstract.

• •

Abstract

You must provide an abstract of no more than 500 words. (Some universities require no more than 150 words for a Master's thesis abstract and 350 for a Doctoral one.) If your thesis is written in a language other than English (for example Afrikaans, Zulu) you must provide two abstracts, one in the language of the thesis and the other in English. The abstract in the language of the main text should come before the translation. The layout of the abstract page/s is as follows:

- Abstract (or *Opsomming*) – as a heading
- Registered title of your thesis (do not translate your title)
- Your initials and surname
- Name of degree [MSc/MPhil/PhD, etc.] Thesis [or Minithesis], Department of ..., University of ...
- The text of your abstract
- Date (month and year)

The abstract should provide a crisp summary of what the thesis undertakes and what major conclusions are drawn. Someone reading the abstract must be able to form a fairly accurate impression of what is in the thesis. As the abstract will be reproduced in various registers and publications (such as *Dissertation Abstracts International* or the *National Research Foundation research list*) available on Internet databases and in libraries across the world, it is important that it should be accurate and well written. A good abstract greatly improves the chances of a thesis being referred to by other scholars.

Note the following:

- The abstract should reflect the balance of the thesis. You should not give much weight to what is a relatively minor part of the thesis, and you should give appropriate weight to an important part of your thesis.
- The abstract is written last, when you know exactly what the thesis covers and what its main conclusions are. Being thoroughly familiar with the work, you should be able to write it fluently and quickly.
- The abstract is not a mere outline of the chapter headings – it is a summary of the argument. It should state clearly what the main research question or problem is, the various sub-issues involved, and your main responses to each.
- You may need assistance in translating your abstract. It is your responsibility to ensure that the translation is accurate, fluent, and grammatically correct.

See Appendix 13 for an example.

• •

Declaration

The following declaration must be included. It must stand on a page by itself under the heading "Declaration":

I declare that ... (full title of thesis, in *italics*) is my own work, that it has not been submitted for any degree or examination in any other university, and that all the sources I have used or quoted have been indicated and acknowledged by complete references.

See Appendix 14 for an example.

Full name .... Date ....

Signed ....

• •

Acknowledgements (optional)

An acknowledgements page is optional. If there are any people you would particularly like to thank, or are obliged to thank (some funders require formal acknowledgement in the finished work), you should do so on this page. Your formulation should be understated and dignified. Avoid sentimentality.

• •

Contents

On the contents page(s), list the preliminary and main parts of the thesis, giving the page numbers on which the parts begin. In the case of chapters, give the chapter number and heading in each case. You must ensure that you give the heading of each part *exactly* as it appears in the main text. Attention to this kind of accuracy is a mark of good scholarship and will be noticed by examiners.

Note:

See
Appendix 15
for an
example.

• If you use sub-headings, you may give the main ones under each chapter heading if you think that will be useful to your reader. However, you must avoid your contents page(s) from becoming too cluttered.

• The contents pages are the last pages of your thesis to be typed, as you need to know the page numbers before you can type them.

• •

Lists of acronyms, tables and/or figures (optional)

If your thesis has many acronyms, figures and tables, it assists your reader if you have lists of these in front. In the list of acronyms, arrange them in alphabetical order, with the full name next to each one. The first time you use the acronym in your text, write out the name in full, followed by the acronym in brackets. Subsequent uses refer to the acronym only.

• •

Preface (optional)

You might include a brief preface *if it would help someone to understand the main text*. You might, for instance, provide some information about the context of the writing of the thesis, or some special hazards, or a formulation of the main question or problem (if you do not do so near the start of the main text).

• •

6.1.2 **Comments on the main text**

Your thesis should be clearly structured and its logic must be evident. The "logic" of your text refers to the principles of reasoning, the sequencing of parts, and conceptual connections that you use in the construction of an overall coherent argument. In other words, it is the thread that links the various parts together in response to your main research problem question. The "structure" of your thesis refers to the way in which you organise your arguments and descriptions. In developing your proposal, you have already thought about the best possible way in which to present your work. Analytical, conceptual, and philosophical studies usually differ in their structure from empirical ones, although there may be overlaps, of course. Broadly, there are theses which draw only on literature and others that combine literature and empirical research.

A theoretical thesis

Theses of this kind usually address academic problems of a historical, semantic, social, ethical, literary, policy-related, legal, metatheoretical or conceptual kind. Most analytical or philosophical theses require at least some conceptual clarification, that is, an attempt to articulate the meaning of key concepts and their relationship to other concepts. These kinds of theses draw only on the literature and typically do not involve doing empirical research. Of course, appeals to written-up case studies and written-up empirical findings are part of the "literature". Instead of having a chapter called "Literature review", these theses draw on the literature (primary and secondary) throughout the development of the thesis discussion and argument.

The aim of the thesis can be one, or more likely a combination, of the following:

- A contextualisation of the problem, that is, an analysis of the circumstances in which the problem or ideas arose
- An attempt to draw links within an existing body of knowledge
- A discussion of the relevance or significance of the research question
- An investigation of the implications of a particular approach or interpretation
- An investigation of the appropriateness of a particular approach or interpretation to the chosen context
- An attempt to trace the development of a particular debate or issue in the literature
- An evaluation (or comparison) of different arguments supporting a particular position
- A comparison and contrast between different research findings in the literature
- An integration of disparate findings as found in the literature
- An identification of a trend (or silence) within the various positions offered on a particular issue
- A clarification of policy and discussion of (possible) consequences of policy implementation
- A clarification of meanings of key concepts as interpreted by various authors
- An examination of the premises, claims and logical construction of central arguments in the discipline.

85

Depending, of course, on the purpose of your thesis and the main research problem question that it tries to answer, chapters in theoretical theses may be structured in a variety of ways. Whatever the structure you decide on, it must have a *logical thread* that binds all the chapters together, that is, there must be some organising principle that guides the sequence of your chapters.

- Chapter sequence may be from general to more specific aspects of the issue.
- Chapter sequence may trace a historical development, that is, move from past to present.
- Chapter sequence may move from international to national to local concerns.
- Each chapter may focus on a specific aspect, position, interpretation, or concept of the overall issue, with the final chapter drawing all the threads together.
- Chapters may follow a formal logical structure, with each chapter building a premise towards an argued-for conclusion.
- Chapter sequence may start with simple, fundamental or uncontested claims/ interpretations and move towards ever-increasing complexity or contestation of the claim/interpretation.
- Chapter sequence may start with a present position, and subsequent chapters may explore possible consequences or implications of that position for the future or for other contexts.

See sections 2.12 and 4.1, Appendix 3.1, and Appendix 15.

How you develop a systematic discussion of your thesis problem is something you will have worked out in your research proposal and chapter outline.

A thesis that combines literature and empirical research

If you are writing a thesis that will involve empirical research, your chapters – whatever sequence or structure you follow – must be organised under appropriate thematic chapter headings and sub-headings. You may, of course, draw on the literature throughout the thesis, as fits the purpose of the particular chapter or the particular sub-section. I have already noted that the rigid "Introduction – Literature review – Methodology – Findings and analysis – Recommendations" template does not allow for much creativity and is therefore increasingly being modified to be much more flexible and responsive to the particular issue that is being researched. However, many universities and departments still insist that students follow this template. In order to assist those students who are required to adhere to this format, there is a list below of the sub-sections within each chapter. But bear in mind that much of contemporary research is not necessarily structured under these headings.

For more innovative structures of empirical studies, see the Petersen example in section 2.12.2 and Appendix 15.2.

The following outline has been adapted from Mouton (2001: 122–126):

Chapter 1 – Introduction

- This is often a reformulation of your research proposal. Start off by saying what the main idea for the thesis is, and why it is an interesting or relevant issue. Give a background to the study, saying how you came to decide on the topic and justifying its academic investigation.

- Show how the topic arises from the reading you have done. First, give a broad overview of the main trends in the academic conversation and then focus more specifically on the aspects your thesis will address.
- Articulate the main research problem that drives your investigation. State what the main academic aims of your thesis are by saying clearly what research *question* the thesis hopes to answer or hypothesis it aims to investigate.
- Give a rough indication of your research framework (what will be your main assumptions, the main perspective from which you will investigate the thesis topic), as well as your research design (what is the structure of your investigation, the steps you plan to follow) and the methodology you'll use.
- Conclude the chapter by giving an outline of each chapter, the main issues each one will address and the logic of the sequence of the chapters.

Chapter 2 – Literature review/theoretical framework

- Begin by indicating the boundaries of your literature review. You can't review everything, so indicate how you decided to limit the field and state your starting assumptions.

See also section 4.1.

- All research has key concepts on which the study draws. You will need to clarify the way in which these concepts are used as early on in the study as possible.
- Discuss the literature you have consulted in an organised and structured manner by grouping readings together under suitable themes. Avoid mere repetition or lists of points. Remember, you need to give your reader a succinct idea of trends in the academic debates, what the various authors are saying, and how the various inputs contribute to the development of the issue you are investigating.
- End your chapter with an overview of the main points that have emerged from your literature review. This is an important step because it will provide the basis for the development of your research. Here you should set up the conceptual tools or categories for the analysis of your findings later on in the thesis.

Chapter 3 – Research design and methods

- State clearly what your research hypothesis is (where applicable), the key concepts and variables.

See also sections 2.10 and 4.2.

- Say what research instruments you will use and discuss what the literature has to say about these (various uses, structures, possible limitations, etc.).
- Discuss why these research methods are appropriate for your thesis purposes – justify your choice. (Include actual copies of questionnaires or observation lists as appendices where applicable.)
- Explain your unit of analysis, the sample design, sampling techniques used and the criteria in terms of which you selected the sample.

87

- Give full details of your data collection processes (times, dates, techniques used, etc.) and any problems you may have had. Give the reasons why you chose to use these processes.

 Van Schaik Publishers

- Describe your data-verification measures. How have you tried to reduce the risk of error?
- Clarify your plan for data analysis. Also describe the data-editing and data-coding procedures you have chosen.
- End the chapter with a discussion of the possible limitations and gaps in the data.

Chapter 4 – Results: presentation and discussion

- Describe the actual sample and its characteristics.
- Describe and summarise the main results.
- Analyse in terms of the categories and conceptual tools you have already signalled in Chapter 2. Discuss and try to explain the main trends, patterns, similarities and differences that may have emerged.
- End the chapter by summarising the main results – positive and negative.

Chapter 5 – Conclusions

- Discuss how the main issues in the literature (Chapter 2) link up with the key findings in Chapter 4.
- Discuss contradictions, gaps, surprise findings and deviations, and suggest possible reasons for these.
- Summarise your argument: state the main problem question your thesis addresses and the main response to it.
- Identify aspects that need further research. (I think it inappropriate for a modest Master's thesis – by its very nature a limited investigation – to make grandiose "recommendations". Rather, be academically more circumspect and suggest "Issues for further research".)
- If appropriate, show how your findings contribute to the general academic conversation.
- Discuss some of the possible implications of your study, either for policy or practice.

Chapter format

Of course, not only must your overall thesis be systematically and logically organised, but each chapter must also follow a structured development. Depending on your purpose, your organisation of headings and sub-headings will differ, but there are some general hints you might want to note when it comes to structuring a chapter.

- Each chapter normally begins by orientating your reader: introduce the main topic of the chapter, the main sections, the way it links to the main research question and, where applicable, how it links with the previous chapter.
- This is followed by the main body of the chapter, organised logically and economically. Make sure that your paragraphs and sections cohere logically (that is, they have a thread which connects them, and form a coherent "story"). The

6

use of sub-headings helps to structure your discussion, although you should avoid using too many.

- Conclude the chapter with a summary of the main points that have emerged. Give your reader "signposts" by noting which issues will be pursued at a later stage, and say how the chapter links with the next one.

See Appendix 10 for some practical suggestions for thesis writing.

6.1.3 Comments on referencing and quotation

Acknowledging your sources is essential in academic writing. You are required to indicate where you found the words you have quoted, or arguments and viewpoints you have paraphrased. This is a matter of honesty, and it shows that you have a sophisticated idea of how ideas develop and are related. If you acknowledge all borrowed ideas, you enable your reader to give you full credit for your own thinking. *Academic readers (including examiners) have little patience or sympathy with academic authors who fail to do so.*

To simplify acknowledgment of sources, we use certain referencing conventions. However, there are various referencing conventions. Check with your supervisor which convention would be most appropriate for you to use. Different disciplines have adopted different referencing practices, so make sure that the one you use is appropriate for the field of study. The golden rule of referencing is to be *consistent* in the convention you adopt.

In-text referencing

Here follows an example of one kind of referencing convention. The one adopted by many journals is known as the "Harvard convention". The format is: (Author, year: page/s). When you are referring to an author's argument or quoting directly, you need to acknowledge this in the text. You can insert the reference either in the sentence, or at the end of the sentence, for instance:

> According to Thompson (1997: 85), ideology is a "set of meanings that sustains relations of domination".

> Or

> "Ideology is a set of meanings that sustains relations of domination" (Thompson, 1997: 85).

If a reference already appears within brackets, then place the year in commas instead of double brackets. For example:

> According to the CDC director, kidney disease is unparalleled in this population (see Table IV in Blake, 1999, for distribution patterns).

89

In the final edit of your thesis, check carefully that each reference in the text is in fact noted in your bibliography. Again, this is something examiners look for.

C

Quotations

You must indicate clearly when you use the actual words of others. There are different ways of doing this. Where the quotation is less than (about) three lines long, mark it with double quotation marks. For example,

> Someone who expresses this point very well is Bobbio (1993: 35) in the striking question, "What is the better form of government, the one based on the rule of men or on the rule of the law?"

Where a quotation is more than three lines long, indent the quotation and *do not use quotation marks* to mark it. We call this indented block a block quotation. You may reduce the line spacing in your block quotations to single spacing and/or reduce the font size, but this is not essential. For example,

> The reform of the laws ... can only be gradual ... but the work is in progress, and both parliament and the judges are every year doing something ... towards the forwarding of it. (Ross, 1996: 72)

Where you omit a word or some words from a quotation, mark that with *three* dots. (See the above quotation.) This same rule applies if you leave off part of the beginning or end of a quoted sentence. The full original of the above quote might read:

> The reform of the law as of any country according to these principles, can only be gradual, and it may be long before it is accomplished; but the work is in progress, and both parliament and the judges are every year doing something, and often something not inconsiderable, towards the forwarding of it.

Please note the following:

- For a quotation within a quotation, use single quotation marks. For example:

  > McGhee (1999: 88) concludes that "the response, then, to Hume's remark that 'morality is more properly felt than judged of' is that of course it is judged of as well as felt".

- When quoting, preserve the spelling and punctuation of the original exactly.
- Punctuation marks *go outside* the quotation marks, unless they belong to the quoted material. For example:

  > What did the author mean by "anti-rotaviral"?

  > Bobbio (1993: 35) asks the question, "What is the better form of government, the one based on the rule of men or on the rule of the law?"

- When you quote, you must know why you are quoting; what purpose does it serve in the progress of *your* argument? Is it to illustrate a point you are making?

Is it to show that you are not misinterpreting something? Never quote simply because you do not know what to say. Your text should make it clear why you are quoting.

- When you incorporate a quotation in your text, make sure that you preserve the grammar of your sentence. For example:

> Bobbio (1993: 35) asks whether the better form of government is "one based on the rule of men or on the rule of the law".

See Appendix 18 for useful websites and reference books.

6.1.4 Comments on notes

There are two main ways of organising your notes: either as footnotes or as endnotes. Some disciplines prefer footnotes (that is, notes at the foot of the page), whereas others prefer endnotes. Check with your supervisor which method would be more appropriate for you to use. Endnotes are listed at the end of your thesis, or at the end of each chapter, whichever makes it more convenient to your reader.

In some disciplines, such as History, notes are used extensively for referencing. In many other disciplines, notes are used only to draw the reader's attention to something important which would obstruct the flow of argument if included in the text. In such a case, notes should be used sparingly, since it is irritating to the reader to have to look up notes which add little or nothing to the overall purpose of the chapter. Notes must have a clear purpose in relation to the main argument. Use them to enrich the discussion while keeping the main line of your argument strong and clear.

If you have an endnotes section at the end of your thesis, it should follow the last page of your main text. Number the endnotes to correspond to the numbers you give in your main text. This is something you need to check with great care in the final copies of your thesis before you submit them for examination. Most word-processing packages have a footnotes/endnotes function, which automatically changes the numbers as you make adjustments to the text, but even so you need to check the correspondence between text numbers and note numbers carefully.

Footnotes or endnotes should be marked in the main text by a continuous series of Arabic numerals, either in superscript or in parentheses (but not both), and they should come after any relevant punctuation. For example:

> ... is the origin of this way of thinking.[5]
>
> or
>
> ... is the origin of this way of thinking. (5)

See Appendix 16 for an example of an endnotes page.

Where you have quite a lot of notes, you might begin a new series of numbers for each chapter; where there are only a few, this is not necessary. If you begin a new series of notes for each chapter, this should be appropriately indicated (by giving the chapter headings in the endnotes).

91

Van Schaik Publishers

6.1.5 Comments on the bibliography

There are two kinds of lists your thesis can have: a *Bibliography*, which lists all the books that you have consulted in the writing and development of your thesis (regardless of whether you refer to them directly or quote from them), and *References*, which lists only the books you refer to directly or quote in your thesis. Check with your supervisor which one is more appropriate for your thesis.

The bibliography or references section comes immediately after the endnotes; it is usually the last section of the thesis. The bibliography should not be artificially bloated (no examiner measures the quality of a thesis by the extent of the bibliography); it must provide a complete list of all the works you used in the development and writing of your thesis, even if you do not explicitly refer to them.

While you are working on your thesis, record the correct information about each work you used on index cards, which can then be arranged in alphabetical order. This can save a lot of trouble later. You need to make especially sure that your bibliography/references contains the full details of every work you refer to in your text and notes. This is the kind of thing that examiners check. Alternatively, you can use one of numerous software packages available that organise your references for you. Enter the full reference details of the text (or Internet site) the first time you use it.

The bibliography should be arranged alphabetically, by authors' surnames, and it should provide all publishing information. Pay especial attention to capital letters, full stops, commas, and so on. Recent conventions aim at simplicity – sometimes the brackets around dates are left out, and the use of capital letters is generally restricted to the first word in the sentence and proper nouns. You may want to use a hanging indent (that is, the first line in an entry is flush with the left margin, the rest is indented), and have single spacing.

See Appendix 3 for examples of different bibliographic listings.

Each entry referring to a paper-based text should have the following details:

Author's surname and initials

Year

Title of article/chapter

Name of journal/Title of whole book

Place of publication and publisher (if a book)

Volume and issue number (if a journal)

Page numbers (if a journal article or a chapter in an edited collection)

Each entry referring to an electronic source should have the following details:

Author's surname and initials (if this is available)

Date on which the site was produced or updated

Title of the document

Electronic medium ("Online", "CD-Rom", or "Electronic" if you're not sure)

Web address (URL link)

Date of your accessing the document

The Harvard system uses the following referencing convention:

Chapter in a book:

Asher, J.J. (1982). The total physical response approach. In Blair, R.W. (ed.), *Innovative approaches to language teaching*. Rowley: Newbury House: 54–66.

Journal article:

Baltra, A. (1992). On breaking with tradition: the significance of Terrell's approach. *Canadian modern language review*, 48 (3): 565–593.

Internet (www) reference:

European Union. (1996). *Living and working in the information society: people first*. Brussels: European Commission. [Online]. Available http://www.ipso.cec.be/html [2003, January 28].

Unpublished comments, discussion, seminar:

Hafkin, N. (1996). *Game theory*. Seminar delivered at the School of Economics, University of Cape Town, 29 March.

Unpublished paper/unpublished thesis:

Marincowitz, F. (1998). *Towards an ecological feminist self*. Unpublished Master's thesis. Stellenbosch: University of Stellenbosch.

Newspaper article:

Gumede, W. (1998). Africa's dream trip over reality at telecom show. *Sunday Independent Business*, 16 May: 5.

Book (single author):

Naisbitt, J. (1984). *Megatrends: ten new directions transforming our lives*. New York: Time-Warner.

Book (more than one author):

Richards, J.C. & Rodgers, T.J. (1986). *Approaches and methods in language teaching: a description and analysis*. Cambridge: Cambridge University Press.

Policy document:

See
Appendix 18
for useful
websites and
reference
books.

Department of Education. (1997). *Education White Paper 3: A programme for the transformation of higher education.* Pretoria: Government Gazette No 18207.

General advice about writing and presenting a thesis

7.1 CLARITY OF WRITING AND READER FRIENDLINESS

An irritated reader is likely to read your work with less sympathy than you would like. Don't irritate your readers, and especially your examiners, by making it difficult for them to understand what you are doing, or trying to do.

Show that you are a serious scholar by ensuring that your thesis is presented with care. To repeat the main technical aspects that need serious attention:

* Make sure that the pages are in the correct sequence.
* Make sure the page numbers and headings listed on the contents page correspond to the sections in the text.
* Avoid unorthodox spelling and punctuation.
* Make sure your referencing follows appropriate and consistent conventions.
* Make sure that every reference in your text and notes also appears in your bibliography.
* Stick to simple rules and patterns (referencing, sub-headings, paragraphing, etc.).

Consistency, even at this surface level, signals that you care about your work.

7.1.1 Some basic rules for reader-friendly writing

Write as plainly and clearly as possible

* Clarity is one of the chief virtues of academic writing. You should be definite (are you making a claim here; is this the third, or the fourth, point in your argument?; is this a claim or an argument put forward by someone else? etc.), and you should orientate your reader, explain what you are doing, remind your reader of the main line of argument of your text, and so on.
* Avoid inappropriate jargon, flowery language and unnecessarily long and complex sentences. In general, shorter sentences are preferable to longer ones.
* Throughout your thesis, refer to it as the "thesis". Do not refer to it as "this thesis" or "this dissertation", and avoid referring to it as "this work".
* Sometimes a heavy block of text needs to be divided into separate paragraphs.

95

Van Schaik Publishers

(Make sure, however, that each paragraph is logically coherent, that is, it has a specific focus that links the sentences together.)

Be consistent

- Especially in headings, sub-headings and in introductory, "bridging" and concluding paragraphs, keep your key vocabulary and rhetoric rigorously consistent. If you refer to "three main problems" in an introductory paragraph, don't allow that to become "three main issues" in the concluding paragraph ten pages later.
- Take particular care to be consistent in your use of referencing conventions.
- Ensure consistency of layout: font size and style of headings, sub-headings and numbering system, throughout the thesis.
- Make sure that headings and page numbers that appear in the contents page correspond exactly to those given in the text.

Use every possible device to avoid obscurity

- No examiner of a thesis will think that obscurity and profundity are synonyms. The aim of good academic research is to clarify and illuminate.
- Avoid using pronouns such as "this" or "it" when there is no clear antecedent.

Prefer literal and direct language

- The trend in academic writing is to avoid using passive constructions (for example, "It was found that …" or "The data that was obtained …"). Rather replace them with "I found that …" or "Mbude's data probably indicates that …".
- The academic trend is to write in the first person, thus clearly identifying your position to the reader. For example, "In my discussion of Newton, I shall highlight the three stages of his axiomatic method". To refer to yourself as "the author of the thesis" or "the researcher" is both stylistically awkward and becoming an outdated way of writing. Most of the top international journals in both the natural sciences and social sciences adopt the "first person" convention.
- In general avoid contractions such "I'll" or "don't" – write them out in full as "I shall" and "do not".
- Especially in headings and sub-headings, avoid being "clever" or pretentious. (Wit and humour typically fall flat in theses – in general, prefer the standard and the orthodox to the offbeat.)

Orientate your reader

- Provide your reader with advance notice of what is to come, crisp and succinct summaries of the point reached in the overall argument, and brief "bridging" paragraphs as the argument moves from one phase to the next. (As you proceed, articulate lucidly what you are doing and how it contributes to your overall project in the thesis.)

- Some disciplines encourage the use of sub-headings (probably at least one every two or so pages) to help readers follow the argument, whereas other disciplines avoid them. Check with your supervisor which approach is appropriate for your purposes. You might have two or three levels of sub-headings, distinguished by different typefaces, centring or underlining. The clarity of your text might be improved by introducing a simple numbering system for part of your argument.

7.2 LAYOUT AND NUMBERING

Again, consistency needs to be your watchword. The layout of chapter headings, section headings and sub-headings must be consistent throughout. If someone is typing the thesis for you, you need to discuss the details of the layout of your thesis with your typist before the work begins.

Margins

The pages of your thesis should have (about) a 3 cm margin at the top and bottom and on the right-hand side, and (about) a 4 cm margin on the left-hand side (to allow space for binding).

Spacing

The text line should be one and a half-spacing (some institutions require double spacing – check with your supervisor). Use only one space after a full stop. Paragraphs start flush left and are separated by a double line space.

Page size

A4 size is the usual requirement, printed on one side of the page only.

Font

The norm for the text is 12 point Times Roman. Sometimes block quotes are in reduced font and/or reduced line spacing. Avoid fancy fonts that are difficult to read. You may want to use different font sizes to indicate a hierarchy of headings, with a bigger font for your chapter headings.

Visual layout

Try to ensure that your pages look attractive, neither artificially spread out nor cramped. In paying attention to your layout, you are trying to make it as easy (and pleasant) as possible for someone to read your thesis; you are showing that you take your own efforts seriously.

Alignment

Full justification (i.e. when both the left and right sides are in a straight line)

blocks the text neatly, but it can lead to some words spreading out: If you use left alignment only, you prevent unnaturally big or cramped spaces.

Ordering of headings

You can use different fonts (size, bold, italics, underline) to indicate descending levels of headings, for example:

CHAPTER HEADING
Section heading
Sub-section heading

Or you could use a numbering outline, alternating letters and numbers and/or degrees of indentation to indicate descending levels, for example:

The consistent labelling optimisation problem in Artificial Intelligence
I. Introduction
    A.   Heuristic search techniques
        1.   Constraint propagation
           (a) Variables
           (b) Values
        2.   Forward-checking algorithm
    B.   Logical Theorem generated search
        1.   Set expansions
        2.   Fix-point mappings

Or, again, you could use a numeric outline, for example:

The consistent labelling optimisation problem in Artificial Intelligence
1. Introduction
    1.1 Heuristic search techniques
        1.1.1 Constraint propagation
           1.1.1.1 Variables
           1.1.1.2 Values
        1.1.2 Forward-checking algorithm
    1.2 Logical Theorem generated search
        1.2.1 Set expansions
        1.2.2 Fix-point mappings

However, this often becomes cumbersome if there are too many sub-headings.

7.3 CHOOSING VISUAL AND ORGANISATIONAL AIDS

Most theses written in scientific and technical fields depend heavily on tables and other visual illustrations to depict and compress information. When properly used, visual aids have several important functions: they summarise, emphasise key points, simplify information and reduce narrative length.

Visual and organisational aids can include a tremendous variety of text accompaniments – tables, line graphs, histograms, maps, photographs, drawings, and the like. You need to decide which graphic format would be most appropriate or most informative for your purposes. The following brief table may help you to determine which type of organisational aid would be the most appropriate for your purposes:

To accomplish this:	Choose one of these:
To present exact values, raw data, or data that does not fit into any simple pattern	Table, list
To summarise trends, show interactions between two or more variables, relate data to constants, or emphasise an overall pattern rather than specific measurements	Line graph
To dramatise differences, draw comparisons	Bar graph
To illustrate complex relationships, spatial configurations, pathways, processes, interactions	Diagram
To show sequential processes	Flow chart
To classify information	Table, list, pictograph
To describe parts	Schematic diagram
To describe a process or organisation	Pictograph, flow chart
To compare/contrast	Pictograph, pie chart, bar graph
To describe change of state	Line graph, bar graph
To describe proportions	Pie chart, bar graph
To describe relationships	Table, line graph
To describe causation	Flow chart, pictograph
To describe entire object	Schematic diagram, illustration, photograph
To show the vertical or horizontal hierarchy of an object, idea or organisation	Flow chart, drawing tree, block diagram

Source: Matthews et al. (1996: 33), reprinted by kind permission of the publishers.

7.4 PAGE NUMBERING

Every page of your thesis, from the title page to the final page of your bibliography, is formally numbered, although some pages (for instance the title page, the abstract and the contents page) might not actually show the number.

Two sequences of numbers should be used. The preliminary pages should be numbered with lower-case roman numerals (i, ii, iii, iv, v, ... etc.). The main page numbering, in Arabic numerals (1, 2, 3, ...), begins with the first page of your main text, and runs through to the last page of your bibliography.

• •

7.5 PROOFREADING

Proofreading is extremely important in correcting a printout of your text. Approach the task in a *technical* frame of mind. You are not reading for content or logical connections at this stage. Once the corrections have been done, you will need to proofread a second printout to pick up any missed or further errors.

7.5.1 Proofreading for general features

In proofreading for such aspects, it is a good idea to make yourself a list of what you are going to check, and then to *range across the whole printout* checking those items. In proofreading a printout of your text, you need to check:

- the consistency of the layout and typeface used for chapter and other headings
- that the numbering of the notes corresponds to the reference in your text
- that every reference in the text is listed in the bibliography (ensure that the spelling of surnames and the year of publication are consistent)
- the page numbering
- the accuracy of the contents page
- other *general features* of the presentation of your thesis.

7.5.2 Proofreading for details

You also need to check details of spelling, punctuation, section numbering, paragraphing, that there are not paragraphs or lines left out or repeated, and so on. These aspects of proofreading can be done only by reading through the text (with a copy of the original ready at your side). Computer spell-checkers can assist in your proofreading, but recognise that they are not foolproof. Words that sound the same (for example, "their" and "there") will both be recognised as correct, but you may have used the incorrect version in your text. Also, although many PCs have grammar checks, these are limited in what they can do – they might not pick up concord (a common error) or the incorrect use of tense. So, ultimately, the responsibility rests on you to read through your text very carefully.

Poor proofreading can seriously affect a reader's (or your examiners') response to what might, in other respects, be a fairly competent piece of work. Some theses get sent back by examiners because the proofreading has been so poorly done. Remember it is the *author's,* that is, *your* responsibility to ensure that the version of the thesis delivered for examination is perfect in all technical respects.

Proofreading is different from editing. Editing is concerned with substantial changes to the text, for instance changing the words used, removing some pieces of text, adding new pieces in, and so on. This is sometimes risky – it can shade over into tampering with the sense of the text. You, *the author,* are responsible for composing the text and controlling its sense.

7.6 GRAMMAR

7.6.1 Some basic grammar rules

Apostrophes

Form the possessive singular of nouns by adding 's. Thus: the dog's bone, Benn and Peter's view in this matter. "Mill's view" indicates that the man's name is Mill; "Mills's view" indicates that the man's name is Mills.

In English, 1980s does not have an apostrophe. (In Afrikaans, 1980's is correct.)

Tenses

The use of present or past forms of verbs has a special meaning in scientific papers – it is a way of indicating the status of the scientific work being reported. However, check with your supervisor which tense/s would be most appropriate for your purposes. See also Matthews et al. (1996: 104–106) from which the following examples have been taken. When discussing other people's findings, ideas in literature, and so on, the following general rules might assist you.

Use the present tense:

- for published facts and ideas (for example, "Serological tests *are* commonly used for diagnostic purposes.")
- for repeated events (for example, "Nesting behaviour *has been* studied under many environmental conditions.")
- for referring your reader to figures, tables and graphs (for example, "Antibodies occurred in 11 per cent of our mice, as Table I *indicates.*" – note here that reference to the table is in the present tense, but discussion of the research itself remains in the past tense).

Use the past tense:

- to discuss unpublished results (for example, "The drug *killed* 95 per cent of the M. tuberculosis bacilli.")
- to discuss results that cannot be generalised (for example, "Barber (1980) reported that 28 per cent of the 396 wasps in his study *showed* signs of parasitism.").

Use either past or present tense:

- when referring to an author. However, the part of the sentence that refers to the findings themselves is still given in the present tense (because it is a published work). For example, "Smith (1975) *showed* that streptomycin inhibits growth of the disease organism." or "Jones (1978) *does* not believe that streptomycin is effective."

The materials, method and results sections describe what you did and are normally written in the past tense.

When you link your own findings with the literature, you will use both past and present tenses – this might be the most tricky part to construct in terms of tense use.

Avoiding gender-biased language

It isn't always easy to avoid gender-biased language because English lacks a gender-neutral singular pronoun. And it is often clumsy to use the construction "he or she" or "s/he". Here are some options you may want to use:

- Use a gender-neutral term when speaking generally of other people (for example, instead of "manpower" use "work force" or "personnel").
- If possible, use alternatives in titles or job descriptions (for example, "speaker" or "representative" instead of "spokesman", and "police officer" instead of "policeman").
- Use plural constructions when you can (for example, "Doctors should advise their patients").
- If appropriate, replace "his or her" with "the" or "a" (for example, "Each teacher is responsible for a class register.").

Concord

Many students struggle with this. Here are some brief pointers.

The following singular subjects all take singular verbs:

Each, one, everybody, neither, as in:

Everybody was involved in the project.

Neither is acceptable.

Each of the findings confirms the hypothesis.

Singular collective nouns, as in:

The country was occupied by the Italians for three years.

A large variety is available.

Either ... or/neither ... nor (as singular nouns), as in:

Neither the interview nor the questionnaire was in English.

A singular noun (qualified by a phrase involving a plural noun), as in:

The set of tests shows an increase in the levels of acid.

• •

7.7 SUBMITTING THE THESIS FOR EXAMINATION

Most universities require you to notify your supervisor in writing of your intention to submit your thesis for examination. Intention to submit is usually taken in consultation with your supervisor, who needs to be satisfied that you have fulfilled the necessary requirements. There are also particular deadlines by which you have to submit if you are hoping to graduate at a particular time.

Normally you will need to submit:

• the required number of ring-bound copies. The outside cover should contain only the full (registered) title of your thesis and your initials and surname, and should be marked "Examination copy". Ask your supervisor how many copies you need to submit.
• proof of current registration.
• in some cases, a letter from your supervisor granting permission for you to submit, a letter from a proofreader certifying the technical correctness of the thesis, or a letter certifying the acceptability of the statistics.

Ensure that the copies you deliver for examination are of the highest quality (it is worth getting them commercially copied), as this makes a difference to the way your examiners read your work. Print single-sided pages. Have them properly ring-bound with a plastic cover and a firm backing sheet as the last page. Specifically check the page sequencing in the copies you deliver. Keep the original for your own records and correction.

Note that the examination of a thesis, including the approval of the various committees involved, normally takes two working months. Doctoral theses can take up to three months. Given the time required for examination, committee approval and final corrections, it is necessary that you submit the examination copies of your thesis by the dates stipulated by your university. Unless you deliver your thesis copies by this date, you are unlikely to graduate at the next graduation ceremony.

Check with your supervisor about procedures to follow once the examiners' reports have been approved by the various committees. Usually, there are some changes and corrections that need to be done before you hand in your final copies. (See below.)

See Appendix 17.1 for the administrative process for the submission of theses, and Appendix 17.2 for a checklist for the submission of theses.

• •

7.8 PREPARING FOR THE ORAL EXAMINATION

Some universities require that students, after having submitted the thesis for examination, also have to undergo a "viva" or oral exam. This usually entails

© Van Schaik Publishers

PREPARING THE THESIS FOR EXAMINATION

the internal and external examiners of the thesis and the student. The reason for having an oral exam is that it creates the opportunity for the student to elaborate on some issues that perhaps were not clear in the thesis, and for the examiners to address some concerns and pursue some ideas expressed in the thesis. Despite your feeling nervous and dreading the exam, it can be a stimulating experience. Particularly in the case of there being little or no chance of the thesis failing, the "examination" can be much more of a collegial interchange in which you can discuss your plans for future research and publication possibilities.

The examiners will use the criteria for examination for a thesis as the basis for the kinds of question they will ask. They will have noted points on

See
Appendix 8.

- the scope of the thesis and the interpretation of key concepts
- the literature you have drawn on
- the support you offer for your main responses to the research question
- in the case of empirical studies, the choice of method and sample
- the technical details of referencing, bibliographic layout, correct grammar and spelling, and general presentation of the text.

Prepare yourself to defend your choice for and demonstration of the above. Apart from questions focused on the above aspects of your thesis, expect to be asked the following (Delamont et al. 1997: 151):

- Why did you choose this topic, this method, this sample?
- What would you do differently if you were starting this work now?
- Summarise your main/the most important findings.
- What problems did you face? How did you overcome them?

Examiners will realise that you are nervous. Even so, you'll have to show that you have a firm grasp of the argument that you advance in the thesis. Here are some tips to bear in mind to help you steer between being appropriately confident and modest about your own thesis:

- Have a copy of your thesis, pen and note pad at hand. Use the note pad to jot down the main thrust of the question and points you want to raise, so that you are able to respond to your examiners in a focused and systematic way when it is your turn to talk. Use the copy of the thesis if you need to refer to specific sections or pages.
- Take appropriate time to respond. You don't have to launch into a defence immediately. Organise your noted points quickly before answering the question.
- If you have with you a list of spelling, grammar, typing and referencing errors that you have already spotted in your own thesis since you've handed it in, you'll impress the examiners with your seriousness about delivering a good thesis. (Of course, the examination copy should have been thoroughly proof-read for these errors beforehand, but some inevitably slip through.)
- If you don't understand the question, say so. Politely ask the examiner to repeat

it, or say, "Could you rephrase that, please?"

- Avoid waffling. Address the question directly and elaborate if necessary. Relate the question, if appropriate, to the rest of the thesis.

- Be modest, but not lacking in confidence. If a criticism is raised, respond with something like, "Yes, that is a good point. However, as I explain in Chapter 3, there are also benefits to be gained from the position adopted by the thesis." Or "Yes, I agree with that. In the thesis, however, I have had to limit my discussion to X because of the constrained scope of the topic." Or "I did not know about X at the time and so chose to do Y. However, as I show in Chapter 5, doing Y has developed certain insights that contribute to the general understanding of the main research problem."

- Attend to practical preparations: don't have a heavy meal or alcohol before the oral; wear comfortable and suitable clothing; have a note pad, pencil, and the thesis at hand; make sure you are well in time for the start of the exam; and arrange to meet a friend afterwards with whom to share your post-exam excitement and anxiety.

• •

7.9 FINAL COPIES

After receiving their result, successful candidates are required to make the noted proofreading, editorial and other minor corrections before they print the final version of their thesis. At a stipulated date before the graduation ceremony, you are usually required to do the following:

- Hand in the required number of properly bound (with expensive, usually leather, binding) copies of the final (i.e. corrected) version of the thesis. Your initials and surname and the full (registered) title of your thesis must appear on the outside of the front cover and on the spine. If the registered title is too long for the spine, use an appropriately abbreviated one. Most universities have a Binding Section (normally located in the library) which will do the binding of the final thesis for you.

- Deliver an unbound copy or electronic version (disk, CD-Rom) of the final version of the thesis. Either microfiche copies will be made from this or it will be loaded onto the electronic library. (You will normally have to pay for this.) Check with the relevant administrator what your university requires in this regard.

- Hand in a letter from your supervisor confirming that the necessary changes have been made and that he or she has checked them.

- Take with you proof of registration.

No student will be allowed to graduate unless final copies have been submitted.

Quality is your responsibility. For these final copies, use high-quality copying services, and check once more that the pages are in the right order and that other

© Van Schaik
Publishers

technical details are correct. Your thesis, once passed, will be available in libraries around the country. Do yourself justice and ensure that the presentation is of such high quality that it removes all unnecessary obstacles from your readers.

. . . .

7.9.1 Extra copies

You may want to make at least two further properly bound copies of the final version of your thesis: one to donate to the department (not all departments request this – check) and one for you to keep on your bookshelf (to show your grandchildren in years to come).

. . . .

7.10 A NOTE ABOUT COPYRIGHT

Universities encourage postgraduate students to publish from their thesis where appropriate. The university, however, retains the sole right to print or publish a thesis or any part or summary thereof, or to have it done. Many universities require their students to sign the following cession:

> I ... hereby cede to the University of ... the entire copyright that may in future subsist in any research report or thesis submitted by me to the University in [partial] fulfilment of the requirements for the degree of ... in the department of ...

If you wish to publish your own thesis, or to have it published, you will need to apply to the university for permission. Information on the format (whole/part; soft/hard cover), publisher, date, and so on, must be provided. Permission is usually granted on condition that:

- it is mentioned in the published work that it was submitted to the university in the form of, or as part of, a thesis
- a copy of the published work is submitted to the registrar
- the changes recommended by the examiners of your thesis have been effected in the published work
- you comply with the conditions the university imposes.

> Writing a thesis is a demanding task. However, remember that many before you have managed to complete their thesis, so it is doable. Avoid the thesis becoming too much of a magnum opus. Although you will be writing and rewriting your various drafts (all authors, no matter how widely they have published, do this), there comes a time when the task of revision has to stop. Apply cost-benefit analysis to your work and go for closure.

Appendices

Appendix 1

What a thesis is and what it is not

What a thesis is

In terms of **conception**:

- At Master's level – a small and modest research project, feasible and doable
- A research project informed by a clear research proposal
- A research project which, like all good research projects, has a clear and definite focus
- A research project which poses a research problem or question and tries to answer it
- In addressing the question, it presents and sustains a coherent and informed argument

In terms of **process**:

- A systematic investigation using appropriate research methods and techniques
- A text written in an appropriate academic style, adhering to proper academic conventions

In terms of **learning**:

- At Master's level – a first opportunity to practise being a researcher
- At Master's level – the beginning of an internship in the craft of research
- An indicator of familiarity with the area of research focus
- An indicator of competence regarding the research methods and techniques utilised

In terms of **product**:

- A coherent and well-organised text of required length
- A well-written and well-edited text
- A proofread text (no typing or minor errors)
- A text with a coherent, appropriate and faultless bibliography and demonstrated familiarity with academic conventions
- A text for the supervisor and examiners

In terms of **examination**:

- At Master's level – a text examined by two to three examiners
- At Doctoral level – a text examined by three to four examiners, usually of international repute

- A text examined according to criteria set by the academic community of the university

What a research project is *not*

In terms of **conception**:

- Simply a description of something

In terms of **learning**:

- At Master's level – it need not make an original contribution to the body of knowledge in the subject area

In terms of **process**:

- An ad hoc, haphazard, sloppy exercise of collecting facts and information around an unfocused topic
- A collection of your opinions and anecdotal stories

In terms of **product**:

- A rave about something that angers or irritates you
- A text addressed to policy-makers
- A text addressed to practitioners
- An extended uncritical political pamphlet

In terms of **examination**:

- An assessment of your personal circumstances, or of how hard you have tried – the product is examined

Source: Kruss, G. (undated).

Appendix 2

A checklist for research proposals

Overall evaluation of thesis topic

☐ Is the problem clearly stated and defined?

☐ Is the topic researchable?

☐ Is the research method appropriate for the purpose of the study?

☐ Has the thesis the potential to make a significant contribution to knowledge or practice in the discipline?

☐ Is the scope of the research appropriate for the degree?

☐ Are the limitations of the research area clearly stated?

The title

☐ Is it an accurate reflection of the contents of the proposal?

☐ Is it succinct – not too long, yet descriptive?

☐ Does it contain the key elements of the proposed study?

The abstract (optional)

☐ Does it capture clearly the main problem, main sections and anticipated response to the question or hypothesis?

☐ Is there a clear articulation of the intended research method/s?

☐ Does it reflect understanding of the main debated issues related to this topic?

The aims of the research

☐ Are the aims academically suitable?

☐ Are they clearly specified?

☐ Are they achievable and not too ambitious or vague?

☐ Do they cohere with the title of the thesis?

The rationale for the research

☐ In the case of a Doctoral study, will the research make an original contribution to the field of knowledge?

☐ Has appropriate literature informed the identification of the topic or issue of research?

☐ Is it an issue worth investigating?

The literature review/theoretical framework

☐ Is the literature relevant to the topic?

☐ Does it summarise the main trends in the debates about the topic?

☐ Does it identify some of the justified criticisms in the debates about the topic?

☐ Has the relationship between the topic and previous research been outlined?

☐ Has the main disciplinary approach to research in this field been outlined?

The statement of the problem

☐ Is the problem informed by the debates in the literature?

☐ Does the statement clearly describe the theory, practical problem or previous research from which the problem proceeds logically?

☐ Is it clearly formulated?

☐ Is the problem soluble?

☐ Does the statement capture the focus of the proposed research?

☐ Are the key concepts of the topic and field of study included?

☐ Does the problem relate to something of scientific, theoretical and/or practical significance, or is it aimed at something trivial? Is it a problem worthy of academic investigation?

☐ Is it congruent to the title as well as the aim of the study (does it address the same issue/s)?

The research design

☐ Is it appropriate to the problem in question (conceptual analysis, historical reconstruction, narrative, survey, experiment, case study, etc.). In other words, will the proposed methods help to realise the aims of the research?

☐ Does it justify the choice of methods on the basis of acceptable research practices and conventions?

☐ Does it clearly describe the planned procedures (so that, for empirical studies, it is replicable):

(a) The sampling procedure

(b) The unit of analysis

(c) The interventions (if appropriate) and/or measuring instruments administered to subjects?

☐ Does it specify how validity of data will be controlled (both in terms of data collection and the extent to which the results can be generalised to other groups or situations)?

☐ Is it feasible, is it manageable (not too ambitious or vague)? Is there a clear indication of focus, necessary resources (time, funds, skills, equipment) and availability of access?

☐ Is there a clear indication of how the data will be analysed?

☐ Has ethical clearance been obtained (or is ethical clearance likely to be approved)?

The chapter outline, time line and budget

☐ Does the chapter outline demonstrate a clear idea of the overall logic of the thesis?

☐ Is the time line realistic?

☐ Is the budget realistic?

The proposal

☐ Does it adhere to the stipulated requirements in terms of layout, sections, number of pages, etc.?

☐ Does the proposal show that you have understood the problem and have an informed and clear idea of how you plan to investigate it?

☐ In general, is it relevant, academically appropriate, interesting and doable?

Ten ways to get your proposal turned down

1. Don't follow the directions or guidelines given for your kind of proposal. Omit information that is asked for. Ignore word limits.

2. Ensure that the title has little relationship to the stated aims; and that neither aims nor research problem link to the proposed methods.

3. Produce woolly, ill-defined aims.

4. Make the statement of the central problem or research focus vague, or obscure it with other discussion.

5. Leave the design and methodology implicit; let them guess.

6. Make some mundane task, routine consultancy, or poorly conceptualised data masquerade as a research project.

7. Be unrealistic in what can be achieved with the time and resources you have available.

8. Be either very brief or long-winded and repetitive in your proposal. Rely on weight rather than quality.

9. Make it clear what the findings of your research are going to be, and demonstrate how your ideological stance makes this inevitable.

10. Don't worry about a theoretical or conceptual framework for your research. You want a down-to-earth study that ignores all that fancy stuff.

Source: Centre for Higher and Adult Education, University of Stellenbosch (2001).

Appendix 3

Appendix 3.1: Example of a well-constructed (abridged) proposal for a theoretical study

Towards a measurement of Quality of Life in Occupational Therapy practice: a conceptual study

An adaptation of a Master's proposal by
Modise Mogotsi (2001)

● ●

Keywords

quality of life, occupational therapy, mental health, generic measures, disease-specific measures, subjective indicators, objective indicators, South African health services, training, practice

Abstract

The purpose of this study is to focus on the current international Quality of Life (QOL) conceptual framework that may be relevant to a South African mental health context for occupational therapy (OT) programmes. The aims of the study are: to understand how the QOL concept has been used in academic debates within the context of mental health; to investigate through a literature study what aspects of QOL have been used in South Africa with regard to mental health, and to develop a baseline conceptual framework for measuring quality of life in OT. In this study, I shall describe the competing worldwide interpretations of QOL. Secondly, I shall explain the debate between generic and disease-specific measures of QOL. This I shall follow with a discussion of the dilemma around subjective and objective indicators of QOL. Fourthly, I shall map out the context of South African health practices and of mental health (in OT) in particular. To illustrate parallels between QOL and OT, I shall compare and contrast the two. Finally I shall frame some suggestions geared towards OT training and practices which address QOL issues that could be applicable to South African health care services in general.

Aims

- To set out how the Quality of Life concept has been used in academic debates within the context of mental health
- To investigate through a literature study what aspects of QOL have been used in South Africa with regard to mental health practices
- To develop a baseline conceptual framework for measuring QOL in Occupational Therapy (OT)

Rationale

In South Africa since the 1980s there have been a number of sociological studies that have measured QOL using anchors, such as standard of living and absence of disease/illness. With the relatively steady improvement in living conditions, the focus on QOL shifted towards measuring health conditions in particular. In the 1990s intervention focused on prolonging the lives of individuals who had terminal illnesses. Thus, Quality Adjusted Life Years (QALY) became the predominant preoccupation of health professionals. Due to accelerated processes of intervention, individuals who survived death now wondered about their QOL, since there were social implications and issues around their level of functioning from the different kinds of interventions rendered. That is, mere treatment was just not enough to improve the overall QOL of individuals.

Currently, in South Africa, the Primary Health Care vision statement on health emphasises participation (networking and collaboration) as well as empowerment (skilling and retraining) (Department of Health: PAWC, 2001). It is evident that "Mental health promotion and the provision of mental health services has been a particular area of neglect in the provision of our already compromised health service" (Department of Health: PAWC. 2001, p. 2). The priority interventions planned by the Department of Health in the Western Cape Provincial Government (2001) include a specific focus on the suggested targets for 2001–2004 for consumer (or client) development. It is especially in this last area that I see my thesis making a contribution. It aims to suggest indicators that will enable the measurement of QOL, and in particular of mental health, in people's occupation and/or interaction with their environment. The purpose of the study is to shed some light on how the QOL concept has been interpreted and implemented in our South African Health Services. Also, it will look at how this could be improved or amended. An OT perspective will be used to demonstrate

QOL with regard to the dynamic nature of human interaction with the environment.

Theoretical framework

Quality of Life (QOL) is a concept which stems from a philosophical stance and embraces both the bigger outer world (i.e. the global context) and smaller inner world (i.e. the organism) of human beings (Bond, 1996). From this basic concept Dossa (1992) notes that a whole host of other approaches towards improving quality of life have been developed. For example, Zhan (1992) argues that religion guides people in a specific, restricted way in which they can understand their own worth and that of others, thus determining their own QOL. Sociology, however, made a breakthrough in establishing the type of indicators which directly impact on QOL. As a result it has moved debates about QOL away from religious interpretations to much wider socio-political analyses. Economics too has become central to QOL debates (Jablensky, 1992). Moreover, in health, technological advancement is also seen as contributing to QOL by giving us better life expectancy (Siegrist & Junge, 1989; Wilkinson, Williams, Krekorian, McLees & Fallon, 1992).

The concept of QOL encompasses subjective perceptions and objective measures of well-being and life satisfaction (Gil, 1990). One area of measuring well-being is the way individuals structure their time by engaging in occupations. Occupation here refers to the human being's interaction with the environment, which arises out of an innate urge towards exploration and mastery, as well as the consequent ability to symbolise the essence of human existence and adaptation (Rosenberg, 1995). I shall examine which factors contribute towards significant occupational Quality of Life. Amongst these factors, particular attention will be given to mental health in terms of how the individual interacts with the environment or functions in his or her different spheres of life.

In the thesis, the presumption is that the QOL concept is used widely as an outcome measure of health. In the debates about QOL in mental health, a controversy about generic measures and disease-specific ones has emerged. A further controversy has developed about objective versus subjective indicators of QOL (Greenley, Greenberg & Brown, 1997; Awad, Hogan, Voruganti & Heslegrave, 1995). However, Gough and Thomas (1994) note that each type of measure has reliable, valid and significant information for a particular purpose. Other investigators (Dew & Simmons, 1990) find it even more valuable to use both

© Van Schaik
Publishers

types to complement one another. In the light of a holistic approach, I shall argue that both approaches used in combination could provide us with a richer picture than using only one.

My research focuses on QOL and its links to OT as it forms part of the bigger Health and Welfare structure. Key concepts in this (OT) area are occupation, occupational performance, independence, activities of daily living (and self-maintenance), work, leisure, play, education and communication activities (or lifeskills). Relevant OT adjustment models will be addressed, i.e. Adaptation Through Occupation (Reed, 1980), Adaptive Skills (Mosey, 1970) and the Model of Human Occupation (Kielhofner, 1980).

I shall use Stedman's (1996) framework of measuring QOL in mental health and adapt this for developing QOL indicators in OT. Stedman proposes three levels of use: consumer and carers; clinicians; and management. The first level relates to information about a range of effects in different domains, comparing alternative interventions, monitoring progress and highlighting to planners the impact of disorder. The second level relates to the clinician's monitoring of the patient's progress, monitoring performance, establishing the effectiveness of interventions and comparing cost-effectiveness of interventions. The third level relates to measurement of outcome/effectiveness of service, establishing service needs and priorities and utilising cost-effectiveness to determine services offered. The emphasis throughout these three levels is on specific indicators and monitoring progress.

Definition of terms

- Occupational Quality of Life (OQOL) refers to a dynamic relationship between the individual's perceived satisfaction with his or her life, well-being, and occupation.
- Mental Health refers to a state of continuous coping with life pressures/stresses. Adaptive mechanisms are emphasised and strengthened to maintain or restore balance in human response to environmental stimulus, for better survival.
- Occupational Therapy (OT) refers to skills used by individuals to strike a balance in their performance areas or various life spheres. Engaging in these occupations marks the ability to accomplish and achieve personal meaning, which promotes independence and satisfaction.

Main research question

How can QOL be appropriately measured and evaluated in OT practices in South Africa?

Research design

I will use a purely text-based approach, with the aim of analysing and clarifying some of the key concepts used in OT practices (and legislation). I shall do this by means of conceptual and logical analysis of key debates, and by tracing the historical and legal developments pertaining to QOL. I shall discuss the main academic debates around QOL in OT as reflected in both international and South African literature. For any future fieldwork to be conducted in a new area such as QOL, there needs to be clarity of concepts used. My thesis aims to contribute to this initial and necessary level of research. In this case I will be placing the concepts in the context of OT and South Africa health legislation.

Chapter outline

Chapter 1: Interpretations of the concept of Quality of Life

1.1 Overview of interpretation of Quality of Life – Stedman (1996)

1.2 Specific perspectives on Quality of Life

    1.2.1 Industrialised context (e.g. Canada, USA, Europe, UK and Australia)

    1.2.2 Developing countries (e.g. African context, particularly Southern Africa)

1.3 Historical development of Occupational Therapy – Coetzee (1994) and its link to Quality of Life – Liddle and McKenna (2000)

Chapter 2: Generic and disease-specific measures of Quality of Life

2.1 Introduction: Problem around measurement of Quality of Life – O' Connor (1993) and Epstein (1997)

2.2 Discussion of commonly used general questionnaires: SF-36 (1992), QOLI (1992)

2.3 Disease-specific scales: Sheehan (1996), SAS-R (1998)

2.4 Cross-cultural issues – Fox-Rushby and Parker (1995)

2.5 Problems (or gaps) in measuring Quality of Life in Occupational Therapy

Chapter 3: Objective and subjective indicators of Quality of Life

3.1 Probability concept – Howell (1997)

References (abridged list)

Awad, A.G., Hogan, T.P., Voruganti, L.N. & Heslegrave, R.J. (1995). Patients' subjective experiences on antipsychotic medications: implications for outcome and quality of life. *International Clinical Psychopharmacology, 10(3), 123–132.*

Bond, A.E. (1996). Quality of life for critical care patients: a concept analysis. *American Journal of Critical Care, 5(4), 309–313.*

Coetzee, S. (1994). Theory: Rooting occupational therapy. First Year Lecture series, University of the Western Cape.

Department of Health: Provincial Administration of the Western Cape.

© Van Schaik Publishers

(2001). Discussion document to inform the development of *Mental health strategic notions for the Western Cape Province 2001–2010*.

Dew, M.A. & Simmons, R.G. (1990). The advantage of multiple measures of quality of life. *Scandinavian Journal of Urology and Nephrology, 131, 23–30*.

Dossa, P.A. (1992). Ethnography as narrative discourse: community integration of people with developmental disabilities. *International Journal of Rehabilitation Research, 15(1), 1–4*.

Gil, D.G. (1990). *Unravelling social policy*. Vermont: Schenkman Books Inc.

Gough, I. & Thomas, T. (1994). Why do levels of human welfare vary among nations? *International Journal of Health Services, 24(4), 715–748*.

Greenley, J.R., Greenberg, J.S. & Brown, R. (1997). Measuring quality of life: a new and practical survey instrument. *Social Work, 42(3), 244–254*.

Howell, D.C. (1995). *Fundamentals statistics for the behavioural sciences*, (3rd ed). California: Duxbury Press.

Jablensky, A. (1992). Politics and mental health. *International Journal of Social Psychiatry, 38(1), 24–29*.

Liddle, J. & McKenna, K. (2000). Quality of life: an overview of issues for use in occupational therapy outcome measurement. *Australian Occupational Therapy Journal, 47, 77–85*.

Mogotsi, M., Kaminer, D. & Stein, D.J. (2000). Quality of life in the anxiety disorders. *Harvard Reviews of Psychiatry, 8, 273–282*.

Rosenberg, R. (1995). Health-related quality of life between naturalism and hermeneutics. *Social Science and Medicine, 41(10), 1411–1415*.

Siegrist, J. & Junge, A. (1989). Conceptual and methodological problems in research on the quality of life in clinical medicine. *Social Science and Medicine, 29(3), 463–468*.

Simmons, S. (1994). Quality of life in community mental health care: a review. *International Journal of Nursing, 31(2), 183–193*.

Stedman, T. (1996). Approaches to measuring quality of life and their relevance to mental health. *Australian and New Zealand Journal of Psychiatry, 30, 731–740*.

Wilkinson, G., Williams, B., Krekorian, H., McLees, S. & Fallon, I. (1992). QALYs in mental health: a case study. *Psychological Medicine, 22(3), 725–731*.

Zhan, L. (1992). Quality of life: conceptual and measurement issues. *Journal of Advanced Nursing, 17(7), 795–800*.

Appendix 3.2: Example of a well-constructed (abridged) proposal for a theoretical and empirical investigation

Prioritising the provision of rural infrastructure in South Africa: a case study of two rural districts

An adaptation of a Doctoral proposal by
Samson Muradzikwa (2003)

• •

Keywords

Infrastructure, rural areas, capital-deepening, welfare-enhancing, cost-benefit analysis, Opportunity Value Assessment, roads, equity, KwaZulu-Natal, Eastern Cape

Aims of the thesis

The overall objective of the research is to develop an Opportunity Value Assessment (OVA) instrument, customised to South African conditions, for prioritising in infrastructure development projects and targets. The OVA instrument will provide planners with the ability to influence policy formulation and implementation with respect to using scarce resources in the most efficient, capital-deepening way.

The specific objectives of this research are:

1. To determine and analyse variables that reflect the capital-deepening aspect.
2. To quantify the benefits or value of targeted rural infrastructure provision.

It must be noted that while the goal here is to develop a sophisticated technical instrument, it will not simply be an exercise in autistic algebra manipulation and other such techniques, but will also include some sensible, realistic political economy as a crucial component. This research seeks to develop an instrument that works on the ground, not just in models.

Rationale

The provision of infrastructure is essential in achieving prosperity and a higher quality of life, to which all countries aspire. Infrastructure policy has pervasive impacts: on economic performance; on the shape and structure of human settlements through its impact on physical

121

development and economic structures; on the distribution of income; and on the environment.

Against this background, South Africa and the rest of Africa continue to wrestle with questions about infrastructure. Especially when South Africa is excluded, sub-Saharan Africa lags behind the rest of the world on almost all dimensions of infrastructure development. In 1999, Africa (excluding South Africa) had 171 000 kilometres of paved roads, about 18 per cent less than Poland, and about 10 million telephones – fewer than Brazil – of which half are in South Africa (World Bank, 2000). Infrastructure development in Africa has been constrained by declining levels of public investment on value-adding and welfare-enhancing infrastructure projects. Soaring debt burdens, sluggish economic growth, and increased pressure on African governments to reduce government expenditure, all continue to worsen the backlog in infrastructure that the region faces. Governments have found it easier, and more politically expedient, to reduce spending on capital expenditure (including infrastructure), where they will suffer less of a political backlash than if they were to, say, reduce spending on the public service wage bill! These political considerations have played a role in the state of infrastructure in Africa today.

Rural people's welfare, which reflects their capacity to produce and consume goods and services, depends on the infrastructure to which they have access. However, it is not inevitable that infrastructure provision will contribute to poverty eradication. Poorly designed and inappropriately prioritised infrastructure could result in more costs than benefits.

Conventional cost-benefit analysis (CBA) techniques tend to allocate lower weight to the utilities of the poorer citizens than to those of wealthier citizens, and this could be problematic in a country with a heavily skewed income distribution such as in South Africa. Therefore the policy prescriptions that result from the traditional CBA techniques may be regarded as politically impractical, or ethically improper, or indeed, both (SABITA, 2002).

The World Bank's Highway Design and Maintenance Standards Model (HDM-IV) is an instrument that is used for calculating economically rational approaches to road building and maintenance (World Bank, 2000). Although this model does take into account some of the weaknesses of conventional CBA approaches, it still falls short of explaining to us how to make trade-offs between different infrastructure provisions if revenues are limited and access to capital is constrained.

In effect, when we compare conventional CBA instruments with the HDM-IV device, what we find is on the one hand, narrowly sound economic tools that are difficult to systematically broaden and, on the other hand, an ethically broad tool that does not represent sound economics. A proper OVA instrument must incorporate ways of achieving *both* measurement breadth and economic soundness.

A brief review of the relevant literature

Widespread interest in the social analysis of projects and policies in developing countries dates from the publications of UNIDO (1972) and Little and Mirrlees (1968, 1974). These books, and the extensive literature generated by their proposals, have had a significant impact, not only on the theory of cost-benefit analysis, but also on the way in which people think about development economics. Recent years have seen a considerable extension of the formal theory of project appraisal. While traditional project appraisal methodology was typically applied to industrial and agriculture projects in the formal sector, it is now common to find that governments and various development agencies are concerned with a much wider range of policy issues, including natural resource management, environmental concerns, income stabilisation and programmes to alleviate poverty.

A number of studies on the theoretical approaches to cost-benefit analysis have been conducted. The studies by Zerbe and Dively (1994), Sugden and Williams (1978), Pearce and Nash (1981), Mishan (1988) and Gramlich (1990) provide useful analyses of the key issues related to cost-benefit analysis and their application to various case studies. As this research will highlight, there are specific details and problems of using cost-benefit analysis techniques in developing countries. Such problems are discussed in studies by Dinwiddy and Teal (1996) who combine an introduction to welfare economics, a discussion of project appraisal principles in developing countries, and a survey of cost-benefit problems raised by externalities, risk and the environment. Of particular interest in the Dinwiddy and Teal (1996) study is the use of general equilibrium techniques that extend the analysis beyond the limits of the well-known partial measures of producer and consumer surplus. This research will, in fact, draw on these techniques in developing the OVA instrument customised to South African conditions.

Squire and Van der Tak (1975) synthesised the work of UNIDO and Little and Mirrlees for the World Bank, and a recent book by Brent (1990) extends this analysis. Fitzgerald (1978), Irvin (1978) and Ray

© Van Schaik Publishers

(1984) present various aspects of the subject, and the work by Curry and Weiss (1992) provides a clear account of the way in which the methodology deriving from Little and Mirrlees can be applied. This approach is also reflected in the ODA (1988) guide, which contains useful case studies and welfare-specific applications. Mirrlees (1978) links project appraisal to distribution issues in the sense that proposed projects or policies, although possibly leading to an overall increase in social welfare, can result in some households gaining more than others so that some will actually find themselves worse off.

Specifying and measuring a welfare function has never been easy. Work done by Harberger (1971), Boadway (1974) and Zerbe and Dively (1994) is instructive, as they analyse the cost-benefit methods within a general equilibrium framework. An introduction to computable general equilibrium models can be found in Bergman (1990) and a comprehensive survey by Shoven and Whalley was published in the *Journal of Economic Literature* (1984). These surveys provide a useful link between the formal analysis on general equilibrium models and its empirical application.

Equity and efficiency considerations, aspects of project appraisal that are at the heart of this research, need to be analysed. A discussion on social welfare and distribution weights in the context of developing countries can be found in Layard (1980) and Ray (1984). Little and Mirrlees (1974) advocate the use of consumption weights based on the iso-elastic welfare function. The use of distribution weights is not without its critics. Studies by Harberger (1978) and Browning and Johnson (1984) provide plausible criticisms of the use of such weights.

While there is cumulative evidence to suggest the power and authority of cost-benefit analysis, neither it (CBA) or any other tool for that matter, will inform us how to make trade-offs between, for instance, subsidising schools and investing in roads if revenues are limited and access to capital is constrained. And although this sort of question is fundamentally political in nature (SABITA, 2002), it ought to be economically informed. It is with such issues in mind that this research proposal has been developed.

The research problem

How can the provision of welfare-enhancing and capital-deepening rural infrastructure be prioritised in a context of scarce resources and inherited inequity?

Distinction of terms

The thesis will distinguish between economic and social infrastructure.

Economic infrastructure is that part of an economy's capital stock that produces services that facilitate economic production or serve as inputs in production, or are consumed by households. Economic infrastructure can be divided into three categories: 1) public utilities, including electricity, gas, water, telecommunications, sanitation, sewerage and solid waste disposal; 2) public works, including water catchment in dams and irrigation, and roads; 3) other transport sub-sectors such as railways, roads, seaports and airports.

Social infrastructure also has a critical role to play in improving the quality of life of the population at large. Capital formation that aims to provide services in health, education and recreation has a direct and indirect impact on the quality of life. It directly supports production and trade and indirectly streamlines many aspects of every day. Furthermore, social infrastructure also facilitates investment in human capital, with the associated gains to productivity and real incomes.

Research methodology

Welfare economics raises exceptionally difficult problems at both the conceptual and the practical level. What is 'welfare'? How do we measure it? How do we influence it? A new economic policy, or a particular project in some geographically-defined area, will typically affect the incomes of many individuals and the prices and choices they make. The question is, "How do these changes in prices and incomes translate into changes in 'welfare' "? Is it a relevant consideration that some households affected by public sector intervention are wealthy, while others are poor? What is the sum total impact on 'welfare', given circumstances of skewed distribution of income and/or wealth?

A number of conceptual issues arise in evaluating the costs and benefits of providing infrastructure in impoverished areas. This research will apply some of the aspects of both partial and general equilibrium. The partial equilibrium analysis would typically involve supply and demand diagrams representative of a single competitive market, where the usual market-clearing assumptions are made (Dinwiddy & Teal, 1996). The more extensive information required for general equilibrium analysis can be a serious deterrent to using this approach. However, the recent development of computable general equilibrium (CGE) models have meant that it is now possible to consider a wider approach to the analysis of the impact of proposed

125

policies and projects. Part of the theoretical methodology would be to analyse welfare change within a two-sector general equilibrium model with one representative household, two products, and two factors of production. This research will be applying aspects of general equilibrium theory, drawing on authoritative literature by Boadway (1974), Dinwiddy and Teal (1988), and Bergman (1990).

Issues of the equity and the efficiency of infrastructure provision will arise. In order to reach a decision as to whether a new policy or project is socially desirable or not, it is essential to have a procedure for combining the gains and losses of the various groups affected. In this regard, the research will draw on work by Ray (1984), Little and Mirrlees (1974), and Dinwiddy and Teal (1996) on social welfare and distribution weights in the context of developing economies.

The complexity and scope of this research calls for additional techniques to be employed. The reason for using some game theory is that it can track changing incentives. These will have to be fed back into the OVA structure. This in turn will probably lead to a policy recommendation for stabilising OVA by the use of a capital grant transfer mechanism. For this exercise, the research will draw on the work done by Gintis (1998).

Empirical data: study area and data collection technique

The data used will be based on selected rural municipal infrastructure development in KwaZulu-Natal (KZN) – District DC 28 – and the Eastern Cape (EC) – District M30/1. Both districts are typical examples of poverty-stricken areas in rural South Africa that are starved of basic services and infrastructure. Infrastructure projects in the two districts will be prioritised in terms of the OVA model developed and then welfare effects tested against it. The population of District DC28 is estimated at 1 740 664 persons and in District M30/1 at about 503 874 persons (SIDAP Report, 2002). These areas are large enough to generate statistically significant comparisons.

Integrated Development Plans (IDPs) for these municipal areas have been drawn up in accordance with the Municipal Systems Act. This research will use these IDPs and other municipal strategy documents that are related to rural development and infrastructure projects that the rural communities themselves have already identified.

The research project has institutional buy-in from the South African National Road Agency (SANRA) and the regional municipalities in the area of study, thus assuring access to the necessary data. In addition,

I have been out into the field a couple of times myself, as part of an outreach exercise in building relationships with, and gaining the trust and confidence of, the municipal officials responsible for development projects in their respective areas. I have worked with them in 'firming' up their IDPs and have attended municipal meetings in the area where IDPs and general development projects are discussed. I am confident that the relationship which already exists between myself, SANRA, and the municipal officials on the ground, provides a good platform for conducting the research.

Preliminary bibliography

Bergman, L. (1990). The development of computable equilibrium modelling. In Bergman, L., Jorgenson, D.W. & Zalai, E. (eds), *General equilibrium modelling and economic policy analysis*. Oxford: Basil Blackwell.

Boadway, R.W. (1974). The welfare foundations of cost-benefit analysis. *Economic Journal, 84: 247–268*.

Boadway, R.W. & Bruce, N. (1984). *Welfare Economics*. Oxford: Blackwell.

Brent, R.J. (1990). *Project appraisal for developing countries*. Hemel Hempstead: Harvester Wheatsheaf.

Browning, E.K. & Johnson, W.R. (1984). The trade-off between equality and efficiency. *Journal of Political Economy*, 92: 175–203.

Curry, S. & Weiss, J. (1993). *Project analysis in developing countries*. London: Macmillan.

Dinwiddy, C. & Teal, F. (1988). *The two-sector general equilibrium model: a new approach*. Oxford: Philip Allan.

Dinwiddy, C. & Teal, F. (1996). *Principles of cost-benefit analysis for developing countries*. Cambridge: Cambridge University Press.

Fitzgerald, E.V.K. (1978). *Public sector investment planning for developing countries*. London: Macmillan.

Gramlich, M.E. (1990). *A guide to benefit cost analysis*. 2nd edition. [s.l.]: South Western Publishing.

Harberger, A.C. (1978). The use of distributional weights in social cost-benefit analysis. *Journal of Political Economy*, 86: S87–S120.

Irvin, G. (1978). *Modern cost-benefit methods*. London: Macmillan.

Layard, R. (1980). On the use of distributional weights in social cost-benefit analysis. *Journal of Political Economy*, 88: 1041–1047.

Little, I.M.D. & Mirrlees, J.A. (1968). *Manual of industrial project analysis for developing countries II*. Paris: OECD.

© Van Schaik Publishers

APPENDIX

3

Little, I.M.D. & Mirrlees, J.A. (1974). *Project appraisal and planning for developing countries*. London: Heinemann Educational.

Mirlees, J.A. (1978). Social benefit-cost analysis and the distribution of income. *World Development*, 6(2): 131–138.

Mishan, E.J. (1988). *Cost-benefit analysis: an informal introduction*. 4th edition. London: Unwin Hyman.

Overseas Development Administration (ODA). (1998). *Appraisal of projects in developing countries*. 3rd edition. London: HMSO.

Pearce, D.W. & Nash, C.A. (1981). *The social appraisal of projects*. London: Macmillan.

Ray, A. (1984). Cost-benefit analysis: issues and methodologies. Baltimore, John Hopkins University Press for the World Bank.

SABITA. (2002). *The under-provision and under-capitalisation of road maintenance, rehabilitation and upgrading in South Africa: analysis and measures towards improvement*. A study commissioned by the Southern Africa Bitumen Association, South Africa.

Shoven, J.B. & Whalley, J. (1984). Applied general equilibrium models of taxation and international trade: an introduction and survey. *Journal of Economic Literature*, 22: 1007–1051.

SIDAP. (2002). *Report on the IDPs status quo*. Report, Phase 1, SABITA Infrastructure Development Project, Cape Town, South Africa.

Squire, L. & Van der Tak, H.G. (1975). *Economic analysis of projects*. Baltimore: John Hopkins University.

Sugden, R. & Williams, A. (1978). *The principles of practical cost-benefit analysis*. Oxford: Oxford University Press.

United Nations Industrial Development Organisation (UNIDO). (1972). *Guidelines for project evaluation*. New York: United Nations.

World Bank. (Annual). *World development report*. Oxford University Press for the World Bank.

World Bank. (2000). *Highway design and maintenance standards model*. www.worldbank.org/html/fpd/transport/roads/red_tools/hdm3.htm

Zerbe, R.O. & Dively, D.D. (1994). *Benefit-cost analysis in theory and practice*. New York: HarperCollins.

Appendix 4

Worksheet for choosing and refining your topic

1. Write down a possible topic from:

   A class discussion you had/An issue raised in a coursework module:

   _____
   _____
   _____

   An assignment topic you had to do:

   _____
   _____
   _____

   A journal article/book you read:

   _____
   _____
   _____

   Your personal experience/workplace:

   _____
   _____
   _____

   •

2. Choose one broad topic from the above:

   _____
   _____
   _____

   •

3. Write down two or three keywords from this broad topic:

_____

_____

_____

• •

4. Now adapt this broad topic to link with something specific to your interest, or something specific with regard to place, or time, or age/gender/class/ethnic/race/interest group, or ethical value, or an administrative issue, or a policy aspect, or an innovation, ... Write down another two to three keywords that capture this specific focus.

_____

_____

_____

• •

5. Fill the above keywords into the blocks below (one word per block) and identify the links between the key concepts. Connect them with a pencil line that captures a logical thread.

• •

6. Linking the keywords together in a logical order, formulate a provisional topic:

_____

_____

_____

• •

7. What problem/puzzle/tension/gap is your topic addressing? Merely describing something is not an investigation. Your thesis needs to address a question whose answer, at this stage, is not obvious or clear. Formulate your provisional topic into a problem question. (Start with e.g.: can, ought, is/are, how, why, where, when, who, what, ... etc.)

_____
_____
_____
_____

• •

8. Identify about four academic aims your response to the question will try to pursue. (for example, start with: explore, investigate, identify, determine, interpret, distinguish, analyse, measure, compare, contrast, understand, ascertain, assess, etc.)

• _____
  _____

• _____
  _____

• _____
  _____

• _____
  _____

• •

9. Now write down the provisional title of your thesis:

_____
_____
_____

© Van Schaik Publishers

Appendix 5

List of some theoretical positions

	Position A	Position B	Position C	Position D
Structure and workings of the physical and social world	• real, fixed • regular • systematic • mechanical laws • individual components make up whole	• language structures our knowledge • hidden interests drive practices • hidden laws structure the social and physical world	• real and dynamic world • regular but flexible • social units shape individual components • language shapes our knowledge • practices shape our knowledge • practices develop over time	• no real world • all based on individual interests • flux and chaos • arbitrary connections • no structure
Method of access to and description of that world	• neutral observation • neutral data collection (structured interviews, survey, etc.) • sense data • apply laws to particular case • establish causal links between variables (experimentation, control group, etc.) • measurement	• discourse analysis • analysis of hidden power relations • uncovering of hidden interests (observation, application of laws) • uncovering of hidden causal links/laws	• historical contextualisation • systems approach (experimentation, focus group, etc.) • discourse analysis • action research • reflection on practices and conventions • purposeful observation • flexible interviews	• describe own story • own voice • identity notion • aesthetic response • sensory experience • poetic language use • non-linguistic descriptions (dance, art, etc.) • articulate contradictions
Purpose of knowledge about the world	• to describe • to analyse into components • to predict • to control and improve	• to change power relations • to broaden access to knowledge	• to describe • to analyse inter-relations • to improve and guide practice • to explain	• to enrich experience • to broaden inclusion • to highlight chaos, contradictions, fluidity of world
Kind of knowledge	• neutral • generalisable • universal	• objective • generalisable • universal	• intersubjective • shared agreement based on accepted conventions • some generalisability	• subjective • private • non-judgmental • relativist • not rooted in tradition
Examples of "isms"	positivism, realism, empiricism, modernism	marxism, critical theory, structuralism, hermeneutics	critical theory, interpretivism, phenomenology, constructivism, realism	post-modernism, relativism, social constructivism, empiricism

Note that there is no neat classification of the big "isms" – aspects of empiricism, such as the notion that individual (sense) experience is the base of knowledge, are linked to those of post-modernism and relativism, as well as to positivism.

© Van Schaik Publishers

Appendix 6

Examples of a well-constructed and a less satisfactory literature review

Example of an unsatisfactory review

Sexual harassment has many consequences. Adams, Kottke and Padgitt (1983) found that some women students said they avoided taking a class or working with certain professors because of the risk of harassment. They also found that men and women students reacted differently. Their research was a survey of 1 000 men and women graduate and undergraduate students. Benson and Thomson's study in *Social Problems* (1982) lists many problems created by sexual harassment. In their excellent book, *The Lecherous Professor,* Dziech and Weiner (1990) give a long list of difficulties that victims have suffered.

Researchers study the topic in different ways. Hunter and McClelland (1991) conducted a study of undergraduates at a small liberal arts college. They had a sample of 300 students and students were given multiple vignettes that varied by the reaction of the victim and the situation. Jaschik and Fretz (1991) showed 90 women students at a mideastern university a videotape with a classic example of sexual harassment by a teaching assistant. Before it was labelled as *sexual harassment* few women called it that. When asked whether it was sexual harassment, 98 per cent agreed. Weber-Burdin and Rossi (1982) replicated a previous study on sexual harassment, only they used students at the University of Massachusetts. They had 59 students rate 40 hypothetical situations. Reilley, Carpenter, Dull and Bartlett (1982) conducted a study of 250 female and 150 male undergraduates at the University of California at Santa Barbara. They also had a sample of 52 faculty. Both samples completed a questionnaire in which respondents were presented vignettes of sexual-harassing situations that they were to rate. Popovich et al. (1986) created a nine-item scale of sexual harassment. They studied 209 undergraduates at a medium-sized university in groups of 15 to 25. They found disagreement and confusion among students.

133

Example of a well-constructed review

The victims of sexual harassment suffer a range of consequences, from lowered self-esteem and loss of self-confidence to withdrawal from social interaction, changed career goals, and depression (Adams, Kottke, & Padgitt, 1983; Benson & Thomson, 1982; Dziech & Weiner, 1990). For example, Adams, Kottke, and Padgitt (1983) noted that 13 per cent of women students said they avoided taking a class or working with certain professors because of the risk of harassment.

Research into campus sexual harassment has taken several approaches. In addition to survey research, many have experimented with vignettes or presented hypothetical scenarios (Hunter & McClelland, 1991; Jaschik & Fretz, 1991; Popovich et al. 1987; Reilley, Carpenter, Dull & Bartlett, 1982; Rossi & Anderson, 1982; Valentine-French & Radtke, 1989; Weber-Burdin & Rossi, 1982). Victim verbal responses and situational factors appear to affect whether observers label a behaviour as harassment. There is confusion over the application of a sexual harassment label for inappropriate behaviour. For example, Jaschik and Fretz (1991) found that only 3 per cent of the women students shown a videotape with a classic example of sexual harassment by a teaching assistant initially labelled it as sexual harassment. Instead ,they called it "sexist, rude, unprofessional or demeaning". When asked whether it was sexual harassment, 98 per cent agreed. Roscoe et al. (1987) reported similar labelling difficulties.

Source: Neuman (2000: 461). From Neuman, W.L. *Social research methods: qualitative and quantitative approaches*. 4/e. Published by Allyn & Bacon, Boston, MA. Copyright © 2000 by Pearson Education. Reprinted by kind permission of the publishers.

Appendix 7

Research methods and sources

> All research has a *conceptual* (theoretical) basis – a perspective from which the story is being told, from which the picture is being painted.
>
> All research draws on the literature. (Some research is only theoretical; other research is theoretical and empirical. All empirical findings must be analysed in terms of contextual, conceptual and methodological frameworks.)

The literature

The literature of your discipline is a necessary source on which you draw in order to help you answer your research question.

- *Primary sources* (usually meaning where ideas and findings were first documented): journal articles and scholarly books in which original knowledge is published, research reports, written-up case studies, historical diaries and narratives containing original insights, as well as policies, laws, curricula outlines, etc. Most theses are required to include at least the significant primary sources.
- *Secondary sources*: articles in journals, academic books, and websites (Internet) that report on and discuss the ideas and findings of other authors, review articles, reference books, encyclopaedias, etc.

Experiment

If your research will involve an experiment, you need to address the following questions in your proposal:

- What is the hypothesis?
- What kind of experiment? Why?
- Has it been done before? Where?
- What are the variables? Why these?
- Details of the experiment: Where will it be done? When? How often? Why?
- How many will be involved? Who? What is your sample? How are you going to choose it?
- What preparation will you need to make?
- How will you monitor the experiment?
- What are your assessment criteria and how will you verify the results?

135

Interviews

If your research will involve interviews, you need to address the following questions in your proposal:

- What do you want to find out? What is the purpose of the interviews?
- Who are you going to interview? Why?
- How many are you going to interview? Why this number?
- When/how often are you going to interview? Why?
- What are the criteria for your interview sample? For example, are you going to consider: age, gender, socio-economic levels, status, job position, geographical distribution, institutional affiliations? Why?
- What will be the structure of your interview? Why?
- In what language will you be interviewing? What language will your interviewees respond in? What might be some of the implications?
- How will you ensure confidentiality of information and adherence to ethical principles of research?
- Perceptions of interviewees vs facts. How will you substantiate the responses and test for validity?
- How are you going to analyse your data?
- What are the practical arrangements you will need to make? Access? Time? Equipment? Transcriptions? Costs?

Questionnaires

If your research will involve questionnaires, you need to address the same questions as those for interviews.

Surveys

If your research will involve surveys, you need to address the same questions as those for interviews.

Case studies/Focus groups

If your research will involve case studies or focus groups, you need to address the same questions as those for interviews. Also:

- What will be the limits of your case study?

Observation

If your research will involve observation, you need to address the same questions as those for interviews. Also:

- What will be your observation criteria?
- How will you record these?
- Will you be using triangulation? Why? How will you plan for this?

Statistics/Graphs/Tables

If you are going to draw substantially on statistics in your research, you need to address the following questions in your proposal:

- What is the purpose of including these statistics?
- What are your criteria for analysis?
- Will you be using a computer package? If so, which one? Why?

Appendix 8

Some criteria for judging Master's and Doctoral theses

In reporting on the thesis, examiners are usually asked to respond to the following:

- Is the scope of the thesis clearly defined?
- Is the nature of the topic adequately interpreted?
- Is there evidence of sufficient engagement with the relevant literature?
- Is sufficient command of appropriate techniques of research and analysis demonstrated?
- Is the thesis well structured and coherently argued?
- Does the thesis reveal a command of the formal conventions of scholarship (such as referencing and bibliography)?
- Has the candidate paid adequate attention to linguistic and formal features of presentation such as grammar, style and layout?
- What are the strengths and weaknesses of the thesis?
- Is the thesis successfully proved or pursued?
- In the case of a Master's thesis, do you think that the candidate should be encouraged to proceed to Doctoral study?
- Is the thesis, in whole or in part, suitable for publication?

In reporting on a Doctoral thesis, examiners are asked to comment in addition on the following:

- Does the thesis show proof of original work?
- Is it a distinct contribution to knowledge of and insight into the subject?

There are usually four possible outcomes to the examination process:

1. The thesis, as it stands, passes.
2. The thesis passes, on condition that some minor changes are made to the final copy.
3. The thesis does not pass, but the candidate should be given the chance to rework the thesis substantially and to resubmit it for examination.
4. The thesis fails outright.

Master's minitheses or Master's full theses are usually awarded a percentage mark. (A distinction or *cum laude* mark is usually 75 per cent or above; a pass mark is 50 per cent or above.) Doctoral theses are usually examined on a fail/pass basis only.

© Van Schaik Publishers

Interpretation of percentage marks for Master's minitheses and full theses

85% and over: A truly outstanding distinction; masterly coverage demonstrating advanced levels of understanding, originality and analysis or research (theoretically and/or empirically) over and above that required for other distinction categories below. Worthy of publication as is.

80–85: A strong distinction without reservations: authoritative coverage of relevant material as well as background literature and/or related issues; outstanding presentation in terms of argument, organisation, originality and style. Demonstrates full understanding of subject matter and, at most, minor typographical corrections are required.

75–79: Merits distinction though with some reservations: a more than competent presentation with good organisation and sound critical arguments. Evidence of originality/clear insight/solid depth of understanding. Some minor omissions and/or corrections required.

70–74: Does not merit a distinction, but there is evidence of some originality and flair. The substantive part of the work is competently covered, well organised and lucidly argued. There are omissions or areas where revisions would improve the work.

60–69: Solidly executed, adequate organisation, competent methodology and conclusions adequately drawn. Little originality, if any, but an adequate overall performance. May require some minor revisions.

50–59: No originality, but a competent, albeit pedestrian, review of the literature, a basic understanding of the significance of the issue discussed, and a fairly competent methodology. There may be problems of organisation and expression, of layout and typographical errors, but the work sufficiently exhibits the main features of academic work to pass. Some major revisions may be required.

49 and less: The work is clearly not adequate. It exhibits such a level of disorganisation and incoherence as to be termed incompetent. The work fails to demonstrate familiarity with basic academic conventions of presentation and organisation. A failing mark indicates that it clearly does not pass in its present form. Should there be evidence that the thesis is salvageable, it may be brought into a pass-worthy form if reworked substantially and resubmitted for examination. A mark less than 30 per cent usually indicates that the thesis is so hopeless that it is not able to be salvaged in any way.

Appendix 9

Example of a contract entered into between student and supervisor

On the basis of the clarification of roles and responsibilities, a written contract can be drawn up and should be signed by both parties.

Agreement between

_____

(name of student)pervisor)

and

_____

(name of supervisor)

Regarding postgraduate research for the degree of:

_____

In the Faculty of :

_____

Signed at: _____

On this day of: _____

_____   _____

Student's signature: Supervisor's signature:

1. The candidate has correctly completed an application form and paid the required fees for admission to the programme.
2. The research project will be completed within the time frames allowed for postgraduate study. The candidate will supply necessary progress reports at the required times.
3. Unless otherwise arranged, there will be fixed monthly meetings between candidate and supervisor. More frequent appointments may be made by the candidate as the need arises. These meetings will be arranged by the candidate (the onus is on the candidate to make the appointment).

4. The candidate undertakes to follow the agreed research agenda and to do the specific tasks agreed on or set by the supervisor.

5. The candidate will ensure that all submitted work is written in an acceptable standard of English (or ...) and that it has been properly proofread. Written work to be submitted in printed form with ample spacing.

6. The candidate understands what the consequences of plagiarism and fraud are and agrees to ensure that this is prevented at all times.

7. The candidate undertakes that all research will be ethically conducted.

8. All work submitted by the candidate will be returned within a reasonable time (maximum turnover of one month) by the supervisor, accompanied by written comments on the manuscript.

9. The candidate will strive to prepare an appropriate paper for presentation at a conference.

10. As the work nears completion, the candidate will submit a complete draft of the manuscript. The supervisor reserves the right to suggest changes, even major ones at this stage, as this is the first opportunity he or she will have had to develop a total perspective of the thesis.

11. The candidate and supervisor ensure that all administrative requirements have been met and guidelines have been followed.

12. The candidate will strive to publish a joint article with the supervisor in an accredited journal.

13. The intellectual property rights of the outcome of the research will be determined by the University policy.

Source: Adapted from a contract issued by the Centre for Higher and Adult Education, University of Stellenbosch, 2001.

Appendix 10

Some practical suggestions for thesis writing

Technical writing does not come easily – as the contents of the university library testify. There is no recipe for success, but there are some rules of thumb that you can follow (Delamont et al. 1997: 124, reprinted by kind permission of the authors):

1. Allow yourself time. No one will believe in advance how long analysis, writing-up and checking take.

2. Set yourself deadlines, and hit them with zeal. (And beware the typist who lets you down at the last moment.)

3. Give shape to what you write. There are all sorts of usable models: the hypothetic-deductive theory (theory, prediction, verification); the ritualistic (introduction, review of literature, method, results, discussion); the auto-bio-graphic narrative, and so on. Pick the one most appropriate for the discipline and for your own inclination.

4. Use sub-headings at the side and/or in the centre of the page to structure your text, and include lots of sentences that tell readers where they have been and where they are going next.

5. Method: just say what you did in words that a child would understand. Keep your discussion of methodology, the pros and cons of the various possible methods, to a minimum. Don't feel compelled to report in detail on every-thing you do.

6. Avoid clichés. "Situation", "at this moment in time", "in this regard", and any-thing else said frequently on television, should be avoided.

7. Decide which of your results are the important ones, and give them promi-nent place.

8. Don't allow technicalities to clog up the main text. Put them in appendices.

9. Expect to suffer over the presentation of statistics. Raw data belongs at the back, after the references. The right way of summarising data in the main text may only come to you after weeks of trial and error.

10. Tables should speak for themselves. Don't force your reader to grope around in the main text to discover what your tables mean.

11. Don't pad out your references with works you haven't read.

12. Hack and hack at your own prose. Sentence by sentence, the simplest form is usually the best. At the level of paragraphs and chapters, aim for the sequence that gives you the smoothest flow.

13. Re-examine any piece of jargon. As often as not you will find that it disguises sloppiness. Bear in mind, too, that the educational sciences are interdiscipli-nary. What you write should make sense to any intelligent person, irrespec-tive of his or her particular technical skills.

14. Proofreading your typed version is essential. Ideally, work with a partner and read it aloud, punctuation and all.

• •

This is Spradley's nine-step guide to academic writing (in Delamont et al. 1997: 130):

1. Select an audience.
2. Select your major argument/theme/thesis.
3. Make a list of topics and create an outline.
4. Write a rough draft of each section.
5. Revise the outline and create sub-headings.
6. Edit the rough draft.
7. Write the introduction and conclusion.
8. Reread the manuscript to check that there are enough examples.
9. Write the final draft.

Appendix 11

Blocks to critical engagement

Remember, you must demonstrate your familiarity with the academic debates. Merely telling your own (academically uninformed) story, however passionate and ideologically correct, is no substitute for earnest and rigorous academic engagement.

You can be hindered in your critical engagement with an issue because of certain bad habits. Be aware of these, not only in the author's writing, but also in your own responses to the author:

- **Cultural conditioning**: often signalled by words such as "obviously", "of course", "must". Ask, what are the "givens" or taken-for-granted assumptions that the author makes?

- **Reliance on opinion**: this means to accept blindly on the basis of popular opinion, for example "It is said that ...", "Everybody knows that ...", "It is a well-known fact that ..." (Says who?)

- **Unsupported generalisations**: avoid words such as "all", "every", "none". Real-life results are hardly ever so definite. Rather be more cautious and use words such as "many", "most", "few", etc.

- **Causal fallacies**: a particular danger in impact studies. This is the wrong idea of what causes the event. Just because one event follows another is no evidence that the one caused the other.

- **Attacking the person rather than the idea**: this means to reject an idea as the person forwarding it is someone you don't like because he or she is of a particular gender, race, ethnic group, political party, affiliated to a specific institution, etc.

- **Hasty moral judgment**: to take for granted that something is a good thing. Often signalled by "ought", "should" or "must".

- **Us-them thinking, or either-or thinking**: this makes us believe that there are only two (usually opposing) positions – one is good/true, the other bad/wrong. It often sets up false polarities and ignores other possibilities (for example "capitalism vs socialism"; learner-centred vs teacher-centred"; "progressive vs conservative").

- **Use of labels**: often encourages simplistic thinking (for example, "democratic"). We need to look closely at the assumptions that underlie these labels and the rationale that drives them.

- **Resistance to change**: it can be threatening to let go of preconceived and cherished notions, and of set ways of doing things and of thinking about them.

- **Slanting**: there is nothing wrong with using expressive language, but this in itself cannot be a substitute for argument. Just because someone feels strongly about something doesn't make his or her beliefs true. Expressive language needs to be supported by reasons.

- **Persuasive definitions**: this is a particular form of slanting which takes the following form: you want to criticise something – x (for example, abortion); choose something most people consider bad (for example, murder); define x in terms of that (for example, "abortion is murder of a foetus"); therefore, x is rejected. Critical analysis is often aimed at the level of how a particular concept is interpreted because, from that particular interpretation, other forms of thinking and doing are developed.

© Van Schaik
Publishers

Appendix 12

Example of a title page

THE RELATIONSHIP OF TRUST AND DISTRUST IN LOCAL GOVERNMENT STRUCTURES IN SOUTH AFRICA SINCE 1994

Valerie Natasha Koopman

A minithesis submitted in partial fulfilment of the requirements for the degree of Magister Artium in the Department of Policy Studies, University of the Drakensberg

Supervisor: Dr Rosalie Mkuze

April 2002

Appendix 13

Example of an abstract

ABSTRACT

CARING IN EDUCATION

S.M. Soal

M.Phil. minithesis, Department of Philosophy of Education, University of the Western Cape

In this minithesis, I explore the connection between education and caring. I argue that, although there is a necessary connection between education and caring, caring is neither a sufficient means to, nor end of, education.

I establish the importance and necessity of caring in education through a conceptual exploration of the moral dimensions of both caring and education. The position that I develop maintains that the relationship between caring and education is mediated by a view of morality which sees caring as one virtue amongst many. I distinguish between two senses of caring: caring about and caring for, and argue that both senses of caring are required in educative relationships.

I critically investigate the views of Nel Noddings, put forward in her book, *Caring: a feminine approach to ethics and moral education*, and conclude that she makes the error of treating caring as the sole basis for morality. This leads her to attempt to provide a sufficient account of education in terms of caring.

I then argue that caring about persons and caring about disciplines are jointly necessary requirements for education. They are the criteria in educative relationships which must be satisfied in respect to caring. This also requires that students must come to care about their disciplines if they are to be considered educated. The minithesis is concluded with an exploration of the threats posed to education and the caring within it by institutionalisation.

Keywords

education, caring, ethics, feminist perspective, compassion, minding, disciplines, academic standards, ownership, insitutionalisation

January 1993

147

Appendix 14

Example of a declaration

DECLARATION

I declare that *Transformative architecture: a synthesis of ecological and participatory design* is my own work, that it has not been submitted before for any degree or examination in any other university, and that all the sources I have used or quoted have been indicated and acknowledged as complete references.

Roberta Fowles   November 2003

Signed: _____

Appendix 15

Appendix 15.1: Example of a contents page of a theoretical study

CONTENTS

Source: Adapted from an M.Ed. thesis by Motshudi, 1994.

Appendix 15.2: Example of a contents page of an empirical study

CONTENTS

Source: Adapted from Egan, 1992, and Nelson, 1985.

Appendix 16

Example of an endnotes page

Chapter 4

1 This constitutes one side of the 'ambiguity' of individualism, which is elaborated below.

2 This supports the suggestion I have repeatedly made that in an inclusive democratic perspective, identification with the social collective is an act of autonomous agency.

3 Indeed, as will be seen, Taylor goes further in linking self-interest and authentic individualism in terms of their common underlying aspiration towards authenticity.

4 Noteworthy in this account is the rejection of differences as "purely accidental characteristics imposed by external forces", as part of the universalist departicularisation of the Enlightenment subject. Ironically, contemporary deconstructionists share the view of the accidental nature of difference, but reject any notion of an essentionalist core.

5 This view contrasts sharply with the Marxist perspective in which individual and collective emancipation are interdependent.

6 This brief account of Dewey draws exclusively from Watt's discussion of his ideas on education and democracy (Watt,1989: 100–114). The quotations in the discussion are drawn sequentially from these pages, unless otherwise indicated.

7 The term "subjectivisation" refers to the shift from external collective horizons of significance to individual subjective ones determined by subjective choice. The modern interpretations of freedom and autonomy, therefore, centre on ourselves, and the "ideal of authenticity requires that we discover and articulate our own identity" (Taylor, 1992: 81).

8 This, as will become evident later, has an important implication for education. If the manner, as opposed to the content, of subjectivity is central to authenticity, then this should have some bearing in a school setting in supporting a method that enhances authenticity. This would suggest shifting emphasis from a content-oriented approach to a skills-oriented approach, which fosters a sense of subjective reflection, judgment and confidence, while retaining due regard for the underlying ideal of authenticity within the strivings of each individual.

151

Source: Subotzky, 1998.

Appendix 17

Appendix 17.1: Administrative process for submission of theses

The rules differ from university to university, but most institutions have some regulations pertaining to:

Registration of thesis title

Once your proposal has been accepted, the thesis title (and keywords/phrases) will be registered on the database. This may be either the university database, or even a national one. Should the title of the thesis change, you (and your supervisor) need to write a letter to the relevant committee, recommending the new title (and new keywords if necessary).

Progress reports and annual re-registration

Most departments require students to submit progress reports to their supervisors at regular intervals (some require a report twice a year, others only once a year). In many cases, your re-registration each year may be dependent on your submission of a progress report, so ensure that you submit this as required. No student may submit a thesis for examination or graduate if the registration is not up to date. You must re-register at the beginning of every year of your study. (Some students are under the impression that once they have registered, they need not register again in subsequent years.)

Extension (or suspension) of registration

Some students do not finish within the allotted time. Usually for a Master's thesis this is a maximum of three years, and for a Doctoral thesis five years. However, most universities also allow students to re-register by permission, granted in exceptional cases. If you need to ask for an extension, you'll have to submit a motivated application, supported by a recommendation from your supervisor. If there are circumstances that you know in advance will delay the completion of your thesis (for example you're expecting a baby, you are going to take on a new job, you are going to be away for a substantial part of the year), you can request a suspension of your studies for the year. What this usually entails is that, on the basis of a motivation letter from you and a recommendation from your supervisor, you will not have to register (and pay registration fees) for that year, and that year will not be calculated as part of the three- or five-year maximum granted for studies.

Notification of intention to submit

Most universities require that you submit a letter to the supervisor informing her of your intention to submit the thesis for examination. Find out the deadlines for submission from the university. On the basis of the notification of your intention to submit, the process of appointing examiners for your thesis starts.

Submission of thesis

Find out from your supervisor how many examination copies of the thesis you must hand in. Some universities also require a letter from your supervisor, giving approval for the thesis to be submitted. Other universities require that each thesis have a certificate of having been proofread and, if applicable, having had the statistics checked. Examination copies of theses are usually ring-bound, with a plastic cover and cardboard backing. The cover page clearly indicates "Examination copy".

After you have submitted your thesis, there are a number of administrative steps you can take while waiting for the results: ensure that all your fees are paid up, and that you have no outstanding fines. Students are not allowed to graduate if the administrative requirements of registration – paid-up fees, returned library books, paid-up fines, etc. – are not met.

Examiners' reports and final corrections

You will not be allowed to know the outcome of the examination until the appropriate committee has approved all the reports. This will normally take about two months from submission. Some examiners agree to a copy of the report being made available to you; others don't. Your supervisor will inform you of the outcome, and the recommended changes if there are any. In many cases, examiners recommend that various minor changes be effected before the final copies are submitted. You will need to meet with your supervisor to discuss the list of recommended changes that need to be effected. Quite often there is not much time between the outcome of the examination and the graduation. Since no student may graduate unless a final copy has been submitted, you will need to do the recommended changes smartly and efficiently. Prepare for this by setting aside some time before graduation. Consult your supervisor on when and where the final copies need to be submitted.

Appendix 17.2: Checklist for submission of theses

Item to be checked	Tick

1. Title page

1.1 Does it have the full registered title of the thesis?	
1.2 Are your names correctly and fully noted?	
1.3 Does it have the correct statement for the degree?	
1.4 Is the month and year noted?	
1.5 Is the name of your supervisor listed (optional)?	

2. Keywords

2.1 Are the ten keywords or phrases listed?	

3. Abstract

3.1 Does it have the full registered title of the thesis?	
3.2 Are your initials and surname noted?	
3.3 Is the name of the degree, department and university noted?	
3.4 Does your abstract adhere to the 500-word limit?	
3.5 Is the month and year listed at the end?	

4. Declaration

4.1 Is your declaration worded according to the requirements?	
4.2 Have you signed it?	

5. Acknowledgements

5.1 Are your funders acknowledged?	
5.2 Are your acknowledgements suitably succinct and dignified?	

6. Contents page

6.1 Are all the preliminary sections, chapter numbers, headings and main sub-headings, endnotes, bibliography and appendices listed?	
6.2 Do the entries on the content page correspond exactly to the ones in the text?	
6.3 Are the page numbers listed correctly?	

7. List of acronyms, tables, figures

7.1 Is the list complete and correct?	

8. Chapters

8.1 Does each chapter start on a new page?	
8.2 Do they all have the correct headings?	
8.3 Are the sub-headings clearly indicated (by number or by layout)?	
8.4 Is there consistency across all chapters with regard to the numbering system, visual layout, ordering of headings, referencing conventions, etc.?	

9. Endnotes/footnotes

9.1 Do the numbers of the notes correspond to those in the text?	

10. Bibliography/references

10.1 Is the referencing convention used correctly and consistently?	
10.2 Does each reference have all the required details?	
10.3 Is the list alphabetised?	
10.4 Is every author or document cited in the text included in the list?	
10.5 Do the spelling of the surname and the date correspond to the reference in the list?	

11. Appendices

11.1 Is each one clearly numbered and do these correspond to the references in the text?	
11.2 Are all the appendices referred to in the text included?	

12. Proofreading and editing

12.1 Have you done a full computer spell check?	
12.2 Have you proofread the entire manuscript?	

13. Page numbering

13.1 Have all pages been correctly numbered and sequenced?	

14. Printing

14.1 Do you have the correct number of required copies?	
14.2 Are they properly ring-bound and marked "Examination copy"?	
14.3 Have you adhered to the administrative requirements of your university?	

15. Copies for yourself

15.1 Have you kept a printed copy for yourself for future reference?	
15.2 Do you have an electronic file of your thesis, safely stored, that corresponds exactly to the hard copy?	

Appendix 18

Some useful websites and reference books

Many websites have sections on "research methods" or "research proposal writing". Bear in mind that different institutions and disciplines have different formats and requirements.

Use **information gateways** for selected sets of information and the selection of suitable topics.

SOSIG – Social Science information gateway
http://sosig.ac.uk/
RDN – Resource Discovery Network
http://www.rdn.ac.uk/
PINAKES – for subject-based information gateways
http://www.hw.ac.uk/libWWW/irn/pinakes/pinakes.html
YENZA! – humanities and social sciences
http://www.nrf.ac.za/yenza/

Virtual libraries

BUBL – Bulletin Board for Libraries
http://bubl.ac.uk/link/
IPL – Internet Public Library
http://www.ipl.org/

South African-based search engine

Useful for locating government documents:
ANANZI – http://www.ananzi.com/
http://www.polity.org.za/lists/govsites.html

Proposal writing

National Research Foundation:
http://www.nrf.ac.za/methods/proposals.htm
http://www.nrf.ac.za/methods/reviews.htm
http://www.nrf.ac.za/methods/guide.htm

http://www.nrf.ac.za/yenza/research/proposal.htm
http://trochim.human.cornell.edu/kb/probform.htm
http://www.aas.org/grants/hints.html
http://fdncenter.org/onlib/shortcourse/prop1.html

For thesis writing and practical advice on getting started

http://www.cc.gatech.edu/fac/spencer.rugaber/txt/thesis.html

http://www.phys.unsw.edu.au/~jw/thesis
http://www.sce.carleton.ca/faculty/chinneck/thesis
http://www.wisc.edu/writing/handbook/academicwriting.html

Statistical data analysis

http://www.stats.gla.ac.uk/cti/links_stats/software.html

For workbooks and academic resources for postgraduate programmes

The Cape Town-based Centre for Research and Academic Development
http://www.radct.co.za

For postgraduate training guidelines

http://www.esrc.ac.uk/ptd/guidelns

Professional bodies and learned societies generally offer a great deal of support to a supervisor and postgraduate student in a variety of ways. Their web address should be accessible via a search engine such as: www.google.com

For supervisor training and codes of practice

http://www.cryer.freeserve.co.uk/supervisors.htm#2
http://www.iah.bbsrc.ac.uk/supervisor_training
http://www.bris.ac.uk/Depts/Registrar/TSU/sofgp99 (for code of practice for UK supervisors)
http://www.qaa.ac.uk/public/cop/cop/annex (for quality assurance indicators of supervision – click on "Supervision" some way down the page)
http://www.npc.org.uk/publications/guidelines/research (For a draft code of practice for postgraduate studies – click on "Responsibilities of the supervisor" some way down the page)
http://www.research-councils.ac.uk (for subject-specific supervisory requirements)
http://www.chem.unsw.edu.au/postgrad/models/TOC (for scientific and technological disciplines, select "Supervisor (and co-supervisors)" from a report from the University of New South Wales)

The following sources may be consulted for further information on ethical considerations

Cooper, D.R. and Schindler, P.S. (2001). *Business research methods*. Boston: McGraw-Hill/Irwin. Chapter 5 deals with ethics in business research.
Berg, Z.C. and Theron, A.L. (1999). *Psychology in the work context*. Cape Town: Oxford University Press, 43–47.
Online Ethics Center: http://onlineethics.org/edu/instruct.html
Ethical issues in health sciences: http://www.cwru.edu/med/bioethics/bioethics
Code of conduct for psychologists: http://www.apa.org/ethics/

Human subjects and research ethics:

> http://www.msu.edu/user/respeck1/HumanResearch.html

National Academy of Sciences' booklet on research ethics:

> http://books.nap.edu/catalog/4917.html

Referencing conventions and guidelines

http://www.lib.uct.ac.za/infolit/bibl.htm (lots of examples)
http://www.pubs.asce.org/authors/index.html
http://lisweb.curtin.edu.au/guides/handouts/ (lots of examples)
http://www.lehigh.edu/~inhelp/footnote/ (footnote and citation style guides)
http://condor.bcm.tmc.edu:80/Micro-Immuno/courses/igr/homeric.html (how to avoid plagiarism)
http://www.wisc.edu/writing/handbook/quosampleparaphrase.html

Useful books for research students

Bell, J. (1999). *Doing your research project: a guide for first-time researchers in Education and Social Science*. Buckingham: Open University Press. (Hints about how to approach research, choose appropriate research methods and interpret the evidence)

Cryer, P. (2000). *The research student's guide to success*. Buckingham: Open University Press. (Good advice about how to manage your academic life as a postgraduate student)

Fairbairn, G.J. & Winch, C. (1991). *Reading, writing and reasoning: a guide for students*. Buckingham: Open University Press. (Tips on academic writing and developing coherent trains of thought)

Henning, E., Gravett, S. & Van Rensburg, W. (2002). *Finding your way in academic writing*. Pretoria: Van Schaik Publishers. (Handy hints on planning, constructing and editing your writing)

Leibowitz, B. & Mohamed, Y. (eds). *Routes to writing in Southern Africa*. Cape Town: Silk Road.

Matthews, J.R., Bowen, J.M. & Matthews, R.W. (1996). *Successful scientific writing*. Cambridge: Cambridge University Press. (Excellent resource for students doing research in the natural sciences)

Mouton, J. (2001). *How to succeed in your Master's and Doctoral studies*. Pretoria: Van Schaik Publishers.

Appendix 19

Sources of funding

There are a number of different sources you may want to pursue:

1. Some departments/postgraduate programmes have grant money available for research projects. Consult with the Head of Department or programme convener for information.

2. Most universities have a composite list of bursaries and funds available for postgraduate students. Enquire from your Student Financial Aid Office, or equivalent unit.

3. A number of national bodies also have funds available for postgraduate studies. Visit the following sites:

   Social sciences, humanities, business:
   > http://www.nrf.ac.za/programmeareas/rsf/grants.stm

   Natural sciences, agriculture:
   > http://www.nrf.ac.za/funding/guide/stud.stm

   Health sciences:
   > http://www.mrc.ac.za

4. The Skills Development Act requires employers to assist their employees in further education and training. You could approach your employer to find out if there is any company funding you could access.

5. The National Research Foundation (NRF) supports postgraduate studies. Consult your supervisor about applying for an NRF grant, or contact the NRF at (012) 481 4000 for further details.

6. Many postgraduate students, all over the world, take out student loans from the bank. Most financial institutions have special deals for students that allow you to start paying back only after having graduated.

©Van Schaik
Publishers

REFERENCES

Barry, V.E. (1984). *Invitation to critical thinking.* New York: Holt, Rinehart & Winston.

Centre for Higher and Adult Education. (2001). Course handout on Supervision Practices. University of Stellenbosch.

Conradie, E. (2000). Research Methodology 711/811: Writing a research proposal. Course handout. Cape Town: University of the Western Cape.

Delamont, S., Atkinson, P., & Parry, O. (1997). *Supervising the Ph.D.* Buckingham: Open University Press.

Egan, K. (1992). *Imagination in teaching and learning.* Chicago: University of Chicago Press.

Kruss, G. (undated). Research Proposal class handouts. Unpublished. University of the Western Cape.

Matthews, J.R., Bowen, J.M. & Matthews, R.W. (1996). *Successful scientific writing.* Cambridge: Cambridge University Press.

Mogotsi, M.J. (2001). Towards a measurement of Quality Of Life in Occupational Therapy practice: a conceptual study. M.Sc. proposal in the Faculty of Community and Health Sciences, University of the Western Cape.

Motshudi, T.F. (1994). *Should the study of traditional African medical practices be part of a unitary educational curriculum in South Africa?* M.Ed. thesis, University of the Western Cape.

Mouton, J. (2001). *How to succeed in your Master's and Doctoral studies.* Pretoria: Van Schaik.

Muradzikwa, S. (2003). Prioritising the provision of rural infrastructure in South Africa: a theoretical and empirical investigation. Ph.D. proposal in the School of Economics, University of Cape Town.

National Research Foundation. (2000). *Workbook for first time and inexperienced researchers.* Pretoria: NRF.

Nelson, K. (1985). *Making sense: the acquisition of shared meaning.* London: Academic Press.

Neuman, W.L. (2000). *Social research methods: qualitative and quantitative approaches.* Boston: Allyn & Bacon, 4th edition.

Petersen, Y. (2002). *Internet outreach practices and technology innovation in environmental education: a grounded study.* Ph.D. thesis, University of the Western Cape.

Robson, C. (1994). *Real World research: a resource for social scientists and practitioner-researchers.* Oxford: Blackwell, 2nd edition.

Smith, M. (2000). Support for post-graduate dissertation writing. In Leibowitz, B. & Mohamed, Y. (eds), *Routes to writing in Southern Africa*, Cape Town: Silk Road.

Soal, S.M. (1993). *Caring in education.* M.Phil. thesis, University of the Western Cape.

Subotzky, G. (1998). *Towards inclusive democratic educational theory and practice in South Africa: mediating individualism and collectivism, difference and communality.* Ph.D. thesis, University of the Western Cape.

Taylor, K. (1996). *Writing academic reviews: an academic survival guide.* The Academic Skills Centre, Trent University, Peterborough, Ontario.

The Economist. (2001). The case for brands. 8 September, p. 9.